THE STUDENT BIOLOGIST
EXPLORES ECOLOGY

THE
STUDENT BIOLOGIST
EXPLORES ECOLOGY

By
Leon Dorfman

Illustrated by
Nancy Lou Gahan

RICHARDS ROSEN PRESS, INC.
New York, New York 10010

Published in 1975 by Richards Rosen Press, Inc.
29 East 21st Street, New York, N.Y. 10010

First Edition

Library of Congress Cataloging in Publication Data

Dorfman, Leon.
 The student biologist explores ecology.

 (The Student scientist series)
 Bibliography: p.
 1. Ecology. I. Gahan, Nancy Lou. II. Title.
QH541.D66   574.5   75–14158
ISBN 0–8239–0327–3

Manufactured in the United States of America

About the Author

Leon Dorfman began his career in science as a laboratory specialist, moved up to teacher of biology, and continued rising to the position of biology department chairman. He currently heads the biology and general science department of the William Howard Taft High School in New York City, as an assistant principal of supervision. Mr. Dorfman is especially known for his effective use of chalk-talk cartoons and magic in  the teaching of science. His techniques have been featured at numerous meetings and conventions of science-teacher associations. He has also written and produced a number of assembly shows on science, one of which was presented on the television program *Carousel.*

Having served as coordinator of science curriculum revision for the New York City school system and biology consultant for the New York State Regents examinations, Mr. Dorfman has been actively involved in the improvement of science instruction and the introduction of innovations. Teachers under his supervision pioneered in the introduction of the Blue Version of the BSCS course in biology for qualified 10th-year students and the Ideas and Investigations in Science course for slow 9th-year students. A number of elective courses have been introduced, including a new course in environmental science.

Mr. Dorfman is the author of the sections on Experiments and Investigations for the books *Exploring a Coral Reef* and *Exploring a Fallen Log*. His review book *Summary Sheets in Biology* has been published by the Rhodes School and is in use in its biology classes.

Contents

Introduction

This book is designed to give the student biologist a deeper understanding of the science of ecology than he might get from a 10th-year biology course. The intent is to develop the basic concepts and pictorial models that make up the principles of ecology so that the student is better prepared to understand, appreciate, and enjoy his environment as well as to take steps to prevent its deterioration. The student will be better equipped, if he so desires, to continue beyond this book to learn more descriptive detail and more challenging mathematical models and field techniques. The basic principles of ecology are carefully developed, however, so that the average student is not left frustrated.

All the concepts and ideas of ecology are consistently tied in with the ecosystem so that the student becomes aware of the environment as a complex system made up of many parts working together. He soon realizes that one cannot change one thing in nature without causing other changes. Some of the changes that are taking place in our environment are described in detail in the last chapter, which deals with man and the environment, and an attempt is made to provide possible solutions to environmental problems.

Some of the newer approaches to the study of ecology are described throughout the book, to give the student as well as the layman an appreciation of the methods of science. Many references are made to research projects being conducted by the United States and other nations in the International Biological Program.

THE STUDENT BIOLOGIST
EXPLORES ECOLOGY

I

The Theme of Ecology

In 1970, when the Apollo 13 spaceship lost the use of its fuel cells and its life support systems were threatened, it had to abandon its mission to the moon and return to earth at once. In that same year, Earth Day served notice that our spaceship Earth was also in trouble and its life support systems were being threatened by man. We were face to face with an environmental crisis.

An awakened generation of Americans, sparked by a group of environmental biologists, suddenly became aware of the seriousness of our declining quality of life and joined the movement to save our planet. Most outspoken and prominent among these scientists were men like René J. Dubos of Rockefeller University, Lamont C. Cole of Cornell, Eugene P. Odum of the University of Georgia, Paul R. Ehrlich of Stanford, Kenneth E. F. Watt of the University of California at Davis, and Barry Commoner of Washington University in St. Louis. They pointed out the impending disaster of the population explosion, the destructive effects of mounting pollutants and wastes on our air, water, and landscape, and the squandering of our global resources of productive lands, wildlife, wildernesses, fossil fuels, and mineral ores. They held out one last hope for survival—the science of ecology.

Ecology has become the most talked about subject of the decade. Phrases like no-lead gas, biodegradable, no phosphates, and recycling have become commonplace in our daily vocabulary.

Accounts of environmental problems appear daily in our newspapers and magazines and are discussed on radio and television programs. Politicians refer to ecology in their speeches, and environmental protection agencies have been established in city, state, and federal departments. Ecology kits and games are displayed in stores. You can become active by joining environmental groups like Friends of the Earth, the Wilderness Society, Zero Population Growth, and local groups in your own community. High school ecology clubs can affiliate with the Ecology Council of America and other similar organizations.

Meaning of Ecology

This book is intended both to give you a basic understanding of modern ecology and introduce you to some of the newer developments. But first, let us define some basic terms. Ecology is the study of the interrelationships between living organisms and their environment. The word is derived from the Greek root *oikos,* meaning "house," and was invented in 1869 by the German zoologist Ernst Haeckel to stress the home or environment in which living things function. Any given environment favors the growth of some plants over others. A forest area, for example, is relatively shady. Therefore, types of plants that need sunlight will not develop if their seeds fall here, whereas types of plants that need little sun will do well. There is, therefore, a nonliving, or abiotic, environment made up of physical and chemical factors that affect the ability of an organism to live and reproduce in any location. This environment includes light, moisture, temperature, oxygen supply, substratum such as soil or rock, and available inorganic substances such as minerals; it may also include pollutants and pesticides. Certain kinds of animals will flourish in the forest if they can find shelter and use as food the kinds of plants that grow there. There will, therefore, be an interaction of organisms influencing the survival of one another. These different organisms make up the biotic, or living, factors in the environment. No living organism is independent of other organisms or the nonliving environment. When we refer to all the members of a species inhabiting a given location, we use the word "population." In a beech-maple forest there will be a population of sugar maple trees, a population of white-tailed deer, and populations of other kinds of plants and animals. When we refer to all the plant and animal populations interacting in a given environment, we use the term "community." For example, the plants and animals living in the forest area make up a forest community.

The Ecosystem Concept

The living community and nonliving environment function together as an ecological system called an ecosystem. A pond, a meadow, and a forest are all examples of ecosystems. Even a small aquarium and a sidewalk crack may be considered ecosystems. Regardless of their size, they are all units in which interaction is taking place between nonliving and living components. Systems that cover large geographical areas are called biomes. Terrestrial, or land, biomes include tundras, coniferous forests or taigas, temperate deciduous forests, grasslands, deserts, and tropical rain forests. Aquatic, or water, biomes are represented by oceans, lakes, and rivers.

Since one ecosystem borders on another, no set rules define the boundaries of an ecosystem. The ecologist who is studying a particular ecosystem defines its boundaries. Very often he will find it convenient to use natural boundaries. For example, he might very well use the edge of a lake as the boundary separating the lake from a forest. At other times, he arbitrarily determines the lateral boundaries of an ecosystem. We may consider the world's ecosystems as forming the biosphere, a thin, global envelope that includes the biologically inhabited air, soil, and water.

Ecosystem Components

The trend today in environmental research is the study of whole ecosystems by teams of ecologists. As they investigate the structure and dynamics of an ecosystem, they find it convenient to recognize four components. To help you understand these components, let us refer to Figure 1, which illustrates a meadow ecosystem, the "Meadow Cafeteria" at lunch time. The cafeteria itself represents the first component, called the abiotic environment, and includes the sunlight, air, and soil. The air contains 78 percent of the element nitrogen, 21 percent of the element oxygen, and .04 percent of the compound carbon dioxide. The soil contains mineral particles that supply compounds containing nitrogen, calcium, phosphorus, potassium, and other elements. Fifty percent of the soil volume is pore space, half of which may be filled with water and half with air.

The grass and other green plants in the meadow make up the second component, the producers, so called because they can manufacture their own food. They carry on the process of photosynthesis, by which light energy is used to combine simple, inorganic compounds into complex, organic materials such as carbohydrates, lipids, proteins, and nucleic acids. These compounds then serve as the source of energy and nutrients needed by plants for maintenance, growth, and reproduction.

Our third component is represented by the field mouse, the snake, and the ants and earthworms in the soil. They are called consumers because they are animals that get their energy and nutrients by eating plants or other animals. The mouse is an example of a primary consumer, or herbivore, because it feeds directly on plants. Other examples of herbivores are crickets,

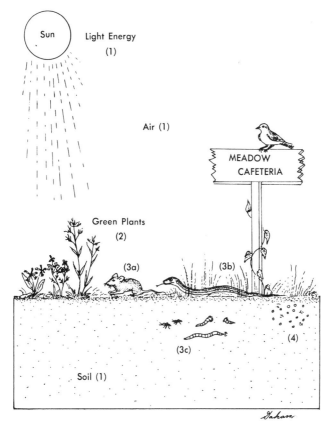

FIG. 1. *The components in a meadow or grassland ecosystem. (1) The abiotic environment. (2) Autotroph or producer. (3) Consumers. (3a) Primary consumer or herbivore. (3b) Secondary consumer or carnivore. (3c) Scavengers. (4) Decomposers—bacteria and fungi.*

grasshoppers, and rabbits. The snake is a secondary consumer, or carnivore, because it feeds on herbivores like the field mouse. Carnivores are predators that catch the animals they eat. Frogs, lizards, and hawks are other examples of carnivores. Some, like the hawk, may prey also on smaller carnivores. The ants and earthworms are called scavengers because they feed on particulate organic matter such as dead leaves in the soil. Some scavengers, for example, vultures, will consume an animal after it has died. Bacteria and fungi make up the fourth component, the decomposers or saprophytes. Decomposers are microscopic consumers that feed on dead plants and animals and break down their complex organic compounds. Some of the products are absorbed for energy and nutrients, and the simplest substances are released for the use of the producers.

One way of separating the components of the biological community is to divide them into autotrophs and heterotrophs. Green plants are autotrophs because they can synthesize their own organic compounds from simple, inorganic substances. Animal consumers and decomposers are heterotrophs because they cannot synthesize organic compounds as the autotrophs do and must eat other organisms or organic matter.

The same ecological components are present and function in much the same way in all ecosystems even though they are populated by very different kinds of organisms. For example, in an ocean ecosystem the autotrophs are microscopic floating plants called phytoplankton (from the Greek *phyton,* which means "plant," and *plankton,* which means "drifting"). There are trillions of these one-celled algae, most of them diatoms, in the surface waters of the ocean where sunlight is most available. Although microscopic, diatoms are capable of producing as much food in a given period of time as are the large, rooted autotrophs on land, given the same quantity of light and minerals. They support a vast array of heterotrophs, the consumers and decomposers of the ocean. Herbivores such as protozoa and tiny shrimplike animals make up the zooplankton, which feed directly on the algae. They are eaten by carnivores such as small fish, which in turn are eaten by larger carnivores such as herring and mackerel. Scavengers such as worms and snails dwell on the bottom sediments and feed on dead organisms falling from above. Also inhabiting the sediments are bacteria and fungi, which function as decomposers. Figure 2 illustrates the components in an ocean ecosystem.

Some Sidelights on Ecosystems

When different species of organisms are found in similar environments but in two different parts of the earth, they are found to be ecologically alike in performing the same function in the ecosystem. For example, the species of grasses in the temperate, semidesert area of Australia are different from those of a similar climatic region in Colorado, but they both function as producers in a grassland ecosystem. In a similar way, the grazing kangaroos of the Australian grasslands are ecological equivalents of grazing bison or cattle on the Colorado grasslands. They live in similar habitats and occupy the same niche in the ecosystem. The term "habitat" refers to the place or environment in which an organism lives, and the term "niche" refers to the role the organism plays in the ecosystem.

Some species may occupy different niches in different habitats or geographical regions. Take man as an example. In some regions, man's food niche is that of a plant eater or herbivore, while in other regions it is that of a meat eater or carnivore. In most cases, man is omnivorous; that is, he eats both plants and animals.

Competition will result when two related species live in the same habitat and use the same food and space. In most such cases, one species, due to differences in reproductive rates, will be successful in eliminating the other. This process of elimination establishes one species per niche in a given habitat and is known as the competition-exclusion principle, or Gause's principle, named after the Russian experimenter who found that two species of paramecium cannot live in the same culture. If two related species are found in the same place, close study reveals that they eat different foods, are active at different times of the day or at different seasons, or in some other way occupy a different niche.

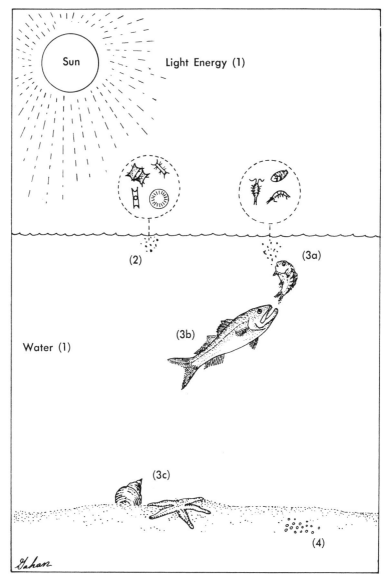

FIG. 2. *The components in an ocean ecosystem. (1) The abiotic environment. (2) Auto-trophic algae of phytoplankton. (3) Consumers. (3a) Herbivores of zooplankton. (3b) Carnivores. (3c) Scavengers. (4) Decomposers.*

When the number of different species of organisms are investigated in ecosystems, it is found that one species will be strongly dominant or present in large numbers, a small number of species will be fairly common, and a large number of species will be quite rare. It is believed that a diversity of species is an advantage to the survival of the ecosystem. The more species that are present, the greater are the possibilities for adaptation to changing conditions. We should consider this principle carefully in planning our agricultural ecosystems. It is risky to depend on only one or a few varieties of wheat because they give the highest yield at the present time. A sudden disease or change in the climate may cause them to become extinct.

The International Biological Program

Ecosystems currently are being researched on a worldwide scale through the International Biological Program (IBP), which opened in July 1967. Participating countries from all parts of the world are investigating two major problems of importance: the ecology of the environment and problems of human adaptability. The ultimate goal of the IBP is to enable man to manage his environment as wisely and efficiently as possible. Future plans include a global network for the environmental monitoring of such variables as temperature, carbon dioxide, pesticides, heavy metals, other pollutants, and possibly the quality of plant and animal species and populations.

In the area of the environment, the United States is investigating the analysis of ecosystems. The first project, the grassland study, will serve as a model for developing research programs on tundras, deciduous forests, deserts, coniferous forests, and tropical rain forests. The program will introduce the new interdisciplinary approach to the study of environmental biology and will use mathematics, systems analysis, and computers. Analog and digital computers will play an important role in using data collected to set up mathematical models of how the grassland system operates. For example, data will be collected in an attempt to describe the pathways between the plant producers and the primary consumers on the Pawnee site, 125,000 acres of grassland near Nunn, Colorado. It is part of the study on small animal diets, one of eleven research projects involving twenty-six investigators. The research team will look at about one hundred species of plants and try to measure for each one the percentages consumed by the 15 to 20 mammalian and 160 insect herbivores. Microscopic examination of digested foods will be used to identify plants eaten by each of these animals. The plants include blue grama grass (often called the "queen of the prairie" because it is so prevalent and hardy), buffalo grass, snakeweed, and two varieties of prickly pear cactus. Included in the mammalian herbivores are two species of jackrabbit, desert cottontail rabbits, black-tailed prairie dogs, ground squirrels, kangaroo rats, pocket mice, and gophers.

Other research projects are under way in the areas of climatology, hydrology, meteorology, photosynthesis, herbage dynamics, large consumers, grassland birds, insects, decomposers, and soil and the nitrogen cycle. These projects will use advanced equipment and techniques, such as neutron probes for measuring soil moisture, capacitance meters to measure changes in the total amount of herbage, microwave radar systems for measuring the amount of evapotranspiration, and low-level remote sensing equipment to detect vegetation and soil differences.

In the study of large animals, one investigation will determine if the native species, such as the bison and the antelope, are more efficient and better adapted to the grasslands than those species—cattle and sheep—imported by man. Alteration of the vegetation by the imported groups will also be studied.

Experimental microwatersheds will be established to study the effects of stresses on grassland. Each microwatershed will be a small area of land surrounded by a concrete curb to collect runoff water from the soil and permit measurement of the volume of water and soil nutrients dissolved in it. Stresses due to large amounts of water, additional nutrients, both water and nutrients, and light, moderate, and heavy grazing will be studied. The objective is to find out if more water or nutrients can change favorably the productivity of the area and lead to better methods of agricultural production without stressing the soil so much that it becomes worthless. The effects on the food web will also be investigated. An important outcome of the IBP is the development of international collaboration in conducting research studies. Canada and Mexico have plans and goals similar to those of the U.S., and twenty other nations are proposing grassland programs. These include Australia, Japan, Poland, the Soviet Union, Israel, and some African countries. As they progress, a worldwide view of grassland ecology will emerge. With this kind of international effort and with the use of systems analysis in the study of ecosystems, it is hoped that answers will be forthcoming to help man solve some of the problems in environmental quality facing him today.

II

The Flow of Energy

How much life and the kind of life that can exist in a given ecosystem depend basically on the rate at which energy flows from one organism to another and the rate at which materials circulate through the system. Energy from the sun may be stored for a while in living matter or used to power life functions, but it is used by organisms only once, is changed into heat, and is radiated away from the ecosystem. On the other hand, water, carbon, nitrogen, and other materials of which living things are composed are used over and over again. This one-way flow of energy and the circulation of materials represent the two basic principles of ecology and apply equally to all environments and all organisms, including man. We will explore the flow of energy in more detail in this chapter and devote the next one to the circulation of materials.

A number of things happen to the energy that comes to the earth from the sun. About 30 percent of it is reflected back into space, mainly from the clouds but some also from the ocean and ground. Most of the remaining sunlight is absorbed, about 50 percent by the ground or oceans and about 20 percent by the atmosphere. This absorbed sunlight heats the earth and is eventually sent back into space as infrared radiation. A very small fraction, about one tenth of 1 percent, is absorbed by green plants in the process of photosynthesis. But as small as this fraction is, the radiant energy of the sun is transformed annually by the world's green plants into about 200 billion tons of the dry organic matter in leaves, stems, roots, fruits, and other plant organs that all heterotrophs, including man, depend on for energy and nutrients. When herbivores graze on plants, much of the food absorbed is used to provide energy for locomotion, digestion, and other functions; only a small part of it becomes flesh. When carnivores consume herbivores, most of the available energy is again used up and little is left for growth. Finally, decomposers use up most of the energy left in dead plants and animals. In all cases, the used energy is changed to heat and is radiated into space. Figure 3 illustrates the energy budget of incoming sunlight and its ultimate conversion to heat.

Photosynthesis and Respiration

Two important processes are involved in the flow of energy: photosynthesis and respiration. In photosynthesis, sunlight is transformed into the chemical energy of carbohydrates, such as sugar, formed by the union of water and carbon dioxide. Oxygen, which is split off from water in one of the reactions, is emitted as a gas. A simplified equation of photosynthesis is shown below:

$$\text{Light Energy} + \text{Water} + \text{Carbon Dioxide} \xrightarrow[\text{Enzymes}]{\text{Chlorophyll}} \text{Chemical Energy in Sugar} + \text{Oxygen}$$

When other organic compounds, such as lipids and proteins, are synthesized from the sugar, they retain its stored chemical energy. It is this energy, the source of power for all living things, that becomes available to them in the process of respiration. Hydrogen, detached as the sugar molecule breaks down, combines with oxygen to form water, and carbon dioxide is left over. Respiration can be represented by the following simplified equation:

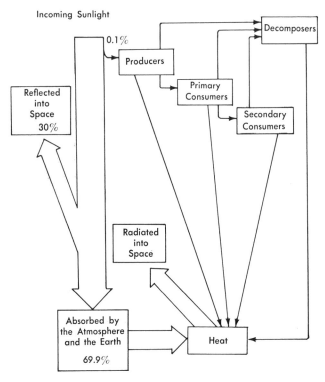

FIG. 3. *The energy budget of incoming sunlight.*

$$\text{Chemical Energy in Sugar} + \text{Oxygen} \xrightarrow{\text{Enzymes}} \text{Energy for Lite} + \text{Water} + \text{Carbon Dioxide}$$

Table 1 conveniently summarizes the essential differences between photosynthesis and respiration.

Food Chains and Food Webs

To study the flow of energy in an ecosystem, an ecologist must first identify the food chains, which show how energy is transferred from one organism to another. This is accomplished by

PHOTOSYNTHESIS COMPARED WITH RESPIRATION

Item	Photosynthesis	Respiration
Organism	Green plants	All plants and animals and most microorganisms
Time of Day	Sunrise to sunset	24 hours a day
Energy	Sunlight *stored* as chemical energy in organic molecules	Energy *released* from organic molecules
Carbon Dioxide and Water	Taken in	Given off
Oxygen	Given off	Taken in

TABLE 1

making direct observations of the feeding habits of animals and also by employing such techniques as analyzing stomach contents and using radioactive isotopes to trace feeding relationships. An example of a food chain in the meadow cafeteria might be:

Grass → Mouse → Snake → Hawk

Each link in the chain represents a feeding, or trophic, level. Grass occupies the first or producer level, the plant-eating mouse occupies the second or primary consumer level, the snake is in the secondary consumer level, and the hawk represents a tertiary consumer. The beginning of the chain is always an autotroph and the end is usually a top carnivore. Hawks and other top carnivores, such as owls and wolves, are not eaten by any larger animals and usually die of old age or accidental injuries. When top carnivores do die, they then serve as food for scavengers and decomposers.

If we look at Figure 4, we can describe other possible food chains in the meadow. For example, grasshoppers and rabbits also feed on grass, and mice can be eaten by hawks as well as by snakes. It is possible to diagram many food chains and find them crossing and intertwining to form a complicated food web. As you trace these food chains within the web, you will discover that some animals occupy different feeding levels in different chains. For example, field sparrows feed on both seeds and grasshoppers and are, therefore, both primary and secondary consumers. Careful study of food chains will also disclose that the animals grow larger as we move up the chain. Although not obvious in the diagram, it is also true that there are fewer of each kind of animal in the community as we proceed up the chain. Since one frog eats many grasshoppers, the number of frogs that live in the meadow must be much smaller than the number of grasshoppers. Each snake, in turn, must be able to hunt over an area that includes many frogs.

Man's Place in the Food Web

Since about 10 pounds of grass are needed to make a pound of beef, a food chain of Grass → Cow → Man is ten times less efficient in the use of solar energy than a food chain of Wheat → Man. An Indian peasant eats 4,000 pounds of grain a year as compared with an American, who eats the equivalent of 2,000 pounds of grain a year—only 150 pounds of it as bread and cereal, the great bulk of it as meat. But the difference between the Indian diet and the American diet is quality as well as quantity. Not only do cereal grains contain much less protein than meat, but their protein also lacks some of the essential amino acids that people need to synthesize human proteins. An Indian living exclusively on a grain diet would eventually suffer from protein deficiency diseases. Nutritionally, that is the reason for paying the ten-for-one price for turning plants into animal proteins.

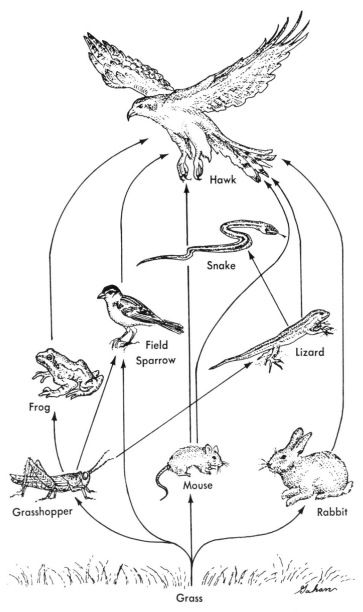

FIG. 4. *A simplified diagram of a food web in a meadow.*

Energy Flow Model

After unraveling the structure of the food web and its food chains, the ecologist then proceeds to estimate the quantity of energy entering and leaving the ecosystem through various trophic levels. Although this is a very difficult task, a few good

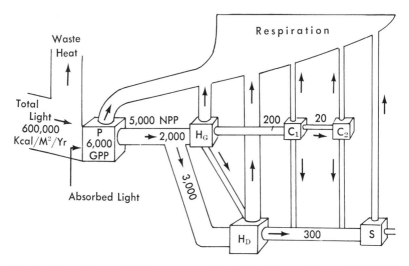

FIG. 5.   *Energy flow model of an ecosystem. (Boxes) Standing crop biomas (KCAL/M²).*
(Pipes) Energy flow in KCAL/M²/YR. (P) Producers. (GPP) Gross primary production.
(NPP) Net primary production. (H_G) Grazing herbivores. (H_D) Herbivore decomposers. (C_1)
Carnivore (secondary consumer). (C_2) Carnivore (tertiary consumer). (S) Scavengers.

examples of such energy budgets do exist and a specific one will be described after we acquaint ourselves with the structure of a general energy flow model, as illustrated in Figure 5. Ecologists use a heat unit called the kilocalorie to describe the amount of energy in a given trophic level, because all forms of energy can be converted to this unit. A kilocalorie may be defined as the amount of heat needed to raise 1,000 grams of water 1° Celsius (centigrade). To determine the number of kilocalories of chemical energy in a given weight of dry organic matter, it is burned completely in the metal compartment of a bomb calorimeter and the rise in temperature of the surrounding water is noted. As a general rule of thumb, 1 gram of dry organic matter is equivalent to 4 kilocalories of energy.

Let us assume that the light energy reaching the ecosystem is 600,000 kilocalories per square meter per year ($kcal/m^2/yr$). Not all of this will become chemical energy for, according to the laws of thermodynamics, in every transformation of energy, a part of this light energy is degraded into heat that is useless for further transfer. Photosynthesis converts only 1 to 5 percent of the sun's energy into organic matter, but this is enough to support all life in the biosphere. If we assume, in our model, that only 1 percent of transformation took place, then 6,000 $kcal/m^2/yr$ will become plant matter. This is known as the Gross Primary Production (GPP). It is shown in the box marked P in Figure 5. The P stands for Producer, and the box represents the standing crop or amount of living material in a population of a given trophic level. The GPP can be expressed as the number of individuals per unit area or the organism mass (biomass) in kcal per unit area. In our figure, no definite amounts are indicated, but the relative sizes of the standing crops of different trophic levels can be inferred from the volumes of the boxes.

Because about 20 percent of the GPP will be used up in plant respiration, 80 percent (or roughly 5,000 $kcal/m^2/yr$), the Net Primary Production (NPP), will be available to the herbivores. In Figure 5, this process is indicated in the pipe leading from the producer box. The pipes represent the energy flow from one trophic level to another. Our model shows two kinds of herbivores, grazing animals (HG) and detritus feeders that feed on plant remains, as well as the decomposers (HD) and scavengers (S). The muscular work of animals demands so much energy that as much as 90 percent of the stored energy will be used up in respiration. Therefore, if they consume 2,000 kcal (3,000 going to the decomposers in our model), not much more than 200 kcal will finally remain as flesh. This will be available to the carnivores (C1), who, in turn, will also use up 90 percent of the energy in respiration and leave only 20 kcal for growth. If the 20 kcal are consumed by tertiary consumers (C2), only 2 kcal will become flesh. Thus, as energy flows through the grazing herbivore chain, the efficiency at each step is only about 10 percent. Energy is similarly lost in respiration in the detritus feeders' pathway.

Each kind of herbivore eats only a small percent of the plant food available to it. The leftover food accumulates along with the herbivore's feces in the surface layer of soil and is broken down slowly by the detritus feeders. It has been estimated that from 40 to 90 percent of the food a grazer eats is never actually absorbed and is voided as feces. Of the food actually assimilated, about 90 percent is used for respiration and there is an additional loss in the form of waste products excreted in the urine. In a similar way, carnivores consume only a proportion of the food available to them and leave over such parts as the bones of herbivores. They do assimilate, however, more of the food than the herbivores do—from 30 to 50 percent and sometimes as much as 75 percent.

In land ecosystems, the detritus pathway is usually greater than the grazing pathway and there is a larger standing crop of plants than of animals. On the other hand, not only is the grazing pathway usually greater in aquatic ecosystems, but there is also a larger standing crop of animals than of plants. This happens because the phytoplankton, even though they multiply rapidly, are eaten so quickly by the zooplankton that their standing crop remains small.

Table 2 describes an actual energy budget worked out by the American ecologist J. M. Teal [1] for a salt marsh in 1962. This

---

[1] Teal, J. M., *Life and Death of the Salt Marsh*. An Atlantic Monthly Book, Little, Brown and Co., Boston, 1969.

QUANTITATIVE STUDY OF THE ENERGY IN A SALT MARSH

Input as light	600,000 kcal/m^2/year
Loss in photosynthesis	563,620 or 93.9%
Gross production	36,380 or 6.1% of light
Producer respiration	28,175 or 77% of gross production
Net production	8,205 kcal/m^2/year
Bacterial respiration	3,890 or 47% of net production
First level consumer respiration	596 or 7% of net production
Second level consumer respiration	48 or 0.6% of net production
Total energy dissipation by consumers	4,534 or 55% of net production
Export	3,671 or 45% of net production

TABLE 2

study of the salt marsh revealed that only 6 percent of the light input was captured in photosynthesis. After losing 77 percent of this chemical bond energy in plant respiration, only 23 percent was stored in plant tissues as net production (8,205 kcal). Of this net production, 47 percent was consumed by fungi and bacteria, 7 percent by herbivores, and about 1 percent by carnivores. This left 45 percent of the net production that was exported with the tides.

World Regions of Productivity

Let us now look at a worldwide view of energy production in different ecosystems, as shown in Figure 6. The values represent the average primary production rate in kilocalories per square meter per day. Three main levels of fertility in the world are illustrated in the figure. Among the most productive regions, ranging between 40 and 100 kcal/m²/dy, are certain shallow water systems such as estuaries and coral reefs together with areas of intensive agriculture such as the year-round culture of sugarcane.

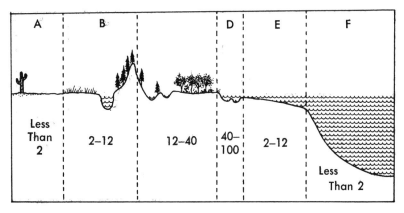

FIG. 6 *World regions of productivity (comparison of gross primary production in different world ecosystems expressed in kilocalories per square meter per day). (A) Deserts. (B) Grasslands, deep lakes, mountain forests, some agriculture. (C) Moist forests, shallow lakes, most agriculture. (D) Estuaries, coral reefs, intensive year-round agriculture. (E) Continental shelf water. (F) Deep sea.*

Included in the second most productive regions, ranging between 2 and 40 kcal/m²/dy, are grasslands, coastal seas, shallow lakes, and ordinary agricultural fields. The least productive regions, ranging from 2 kcal/m²/dy to less, are the deep oceans and land deserts. Generally, terrestrial and aquatic systems can be equally productive if they are supplied with similar quantities of light, water, and nutrients. We can also generalize that a large portion of the earth's surface is in the low production category because of the lack of nutrients in the open sea and the lack of water in the land deserts. In the future, irrigation systems will increase productivity in deserts, and it may be possible to bring up nutrients from the sea bottom.

Increasing Productivity

The most immediate prospects for increasing man's food production lie in measures that provide crops with the best physical and chemical conditions and increase the season of growth. Recent experiments have indicated that the amount of carbon dioxide fixed in photosynthesis can be increased by inhibiting the sudden loss of carbon dioxide that occurs when a plant in the dark is exposed suddenly to light. This process is called photorespiration. Since the carbon dioxide comes from the oxidation of glycolic acid, which starts when light is turned on, one approach to diminishing this reaction is to spray plants with an inhibitor, HPMS (alpha hydroxy pyridine methanesulfonic acid), which is effective in decreasing glycolic acid oxidation. Another approach is to decrease the percent of atmospheric oxygen from 20 percent to 5 percent. However, this requires that the crop be grown in a greenhouse.

A number of synthetic foods have been developed that also have possibilities. One of them, fish protein concentrate (FPC), is made from "trash" fish—fish that are ordinarily scorned by consumers. Another synthetic food, leaf protein concentrate (LPC), is produced by a chemical method that allows humans to digest alfalfa, elephant grass, and the like almost as efficiently as cows do. And a third synthetic food, single cell protein (SCP), is made from microbes grown on petroleum, cheese whey, molasses, or animal droppings. The SCP is treated in order to make the microbes both edible and nutritious.

III

The Cycling of Nutrients

At least thirty of the earth's chemical elements are essential to the growth and development of a living thing. Most of them, however, cannot be used by organisms unless they are combined with other elements in chemical compounds. Hydrogen, for example, is available in water, which has two hydrogen atoms and one oxygen atom in its molecule. These elements and their compounds, called nutrients, move back and forth between the living and nonliving components of the ecosystem in more or less circular paths called cycles. Specifically, they are called biogeochemical cycles because the chemicals are found in the earth's waters, atmosphere, and the rocks and soil in the ground.

Figure 7 illustrates the basic scheme in the cycling of materials. Nutrients, absorbed by producers from the nonliving environment, are assimilated into organic compounds and moved up the food chain to different levels of consumers. After death, the materials are simplified by decomposers and returned to the nonliving environment to be used once again. Microorganisms thus play a key role in replenishing the nutrient pool. Many different kinds of bacteria may be involved in the various stages of decomposition of an organic compound. At each step, a new group of microorganisms will be able to feed on the waste products of the previous group, until the last stage is reached. Many of the details of microbial biology are still not known, and practical applications will not be possible until microbiologists conduct more research in this area.

Because they are essential to living things and are required in relatively large amounts, nine of the elements are called macronutrients. These are listed in Table 3 along with a group of micronu-

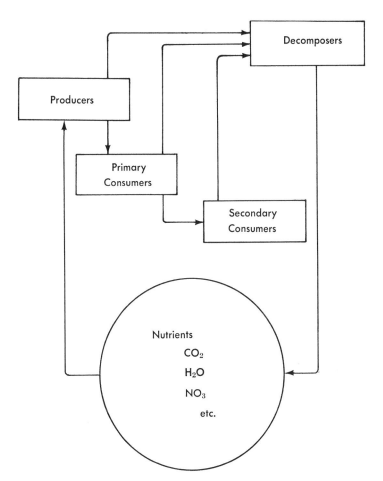

FIG. 7. *The basic scheme in the cycling of nutrients. Nutrients, removed from the abiotic environment by producers, are eventually returned by decomposers.*

trients needed in very minute quantities by green plants. Some of the micronutrients, such as sodium, may serve as macronutrients for certain animals, and additional elements, such as iodine, may be required by other organisms.

NUTRIENTS ESSENTIAL FOR GROWTH AND DEVELOPMENT IN PLANTS

Macronutrients	Micronutrients
Carbon	Iron
Hydrogen	Manganese
Oxygen	Copper
Nitrogen	Zinc
Sulfur	Boron
Phosphorus	Sodium
Potassium	Molybdenum
Calcium	Chlorine
Magnesium	Vanadium
	Cobalt

TABLE 3

The Carbon-Hydrogen-Oxygen Cycle

The first three macronutrients—carbon, hydrogen, and oxygen—which comprise 94 percent of the dry weight of plants, are so important in the processes of photosynthesis and respiration that we can think of them as circulating together in one cycle. Figure 8 illustrates this cycle in a terrestrial system. Water (H_2O) and carbon dioxide (CO_2) are absorbed by producers and synthesized into carbohydrates (CH_2O). Oxygen (O_2), produced as a by-product, is released into the environment.

Aquatic plants take the water they need directly from their surroundings, but land plants obtain their water from the soil. Although water is plentiful, carbon dioxide is limited, making up only 0.03 percent by volume of the atmosphere. Sea water contains about thirty times as much CO_2 in the form of bicarbonate and carbonate ions. Land plants would exhaust the quantity of atmospheric carbon dioxide in a few years if it were not replenished by consumers in the process of respiration. They take in food and oxygen and combine them to release energy for their life activities. Carbon dioxide and water, produced as waste products, are excreted into the environment. Decomposers, the ultimate consumers, carry on respiration in exactly the same way.

The Water Cycle

As indicated in Figure 8, terrestrial consumers excrete water as a vapor (gas). It must go through a longer cycle called the water cycle before it can return to the soil as water. In this cycle, the excreted vapor, together with other water vapor evaporated by the sun's heat from the surfaces of oceans, lakes, and streams, cools off as it rises to high altitudes. It then condenses into tiny droplets of water to form clouds. When the droplets become larger and heavier, they fall as rain, snow, or other forms of precipitation. Most of the precipitation falls back into the oceans, lakes, and streams, and only about one eighth of it falls onto the land. Of the fresh water that does fall on the land, about 30 percent of it will evaporate from various surfaces before it can be used by plants and animals. In hot, dry regions, all the water may be lost in this way. The water that does not evaporate enters the soil and becomes available to plant roots and soil organisms. The plants absorb and assimilate some of the water, but most of the absorbed water reaches the leaves and is lost as water vapor through the microscopic stomatal openings by a process called transpiration. Together, evaporation and transpiration remove 70 percent of the total precipitation falling on land and return it to the atmosphere, where it will condense and fall again as land water.

The remainder of the water entering the soil will seep down until it reaches an impervious layer of rock, along which it will move as groundwater until it reaches an outlet such as a lake,

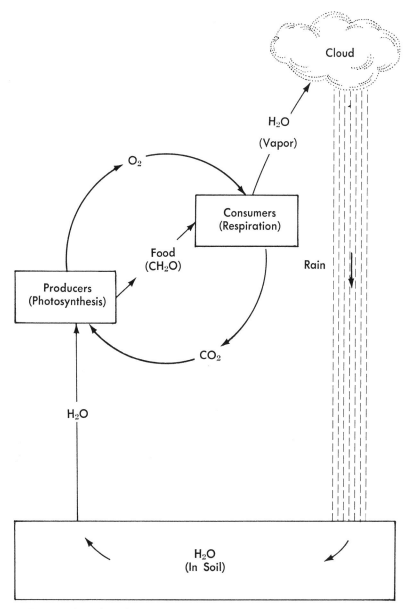

FIG. 8. *The cycling of carbon, hydrogen, and oxygen in a terrestrial environment.*

stream, or ocean. Eventually, all fresh water that falls on land returns to the sea.

The Nitrogen Cycle

Nitrogen, the fourth element on the list of macronutrients, is important in the building of proteins, which make up half the dry weight of living things. In this process, nitrogen is added to sugar molecules to form a number of different amino acids from which proteins are synthesized. Not only are proteins important components of muscles and other parts of the body but, as enzymes, they also regulate the speed of life's chemical reactions. Although four fifths of the atmosphere is nitrogen, plants cannot assimilate it directly but must absorb it from the soil in the form of electrically charged particles called ions, which form when a salt is dissolved in water. For example, a molecule of sodium nitrate will ionize in water into one ion of sodium with a positive charge (Na^+) and one ion of nitrate (NO_3^-) with a single negative charge. These ions move around independently in the soil solution and will enter root hair cells if admitted by the selective action of their membranes.

The nitrogen cycle can be thought of as two separate movements, one involving the atmosphere and the other involving the decay of organisms. In the atmospheric cycle, shown as A in Figure 9, nitrogen can become available to plants through the action of nitrogen-fixing bacteria that live in the roots of legume plants such as beans, peas, and clover. These bacteria stimulate the roots to surround themselves with growths called nodules in which the bacteria fix atmospheric nitrogen into nitrates. These are then used by the legume plants for protein synthesis. In return, the roots provide food and shelter for the bacteria in a helpful relationship called mutualism.

Although this close connection between nitrogen-fixing bacteria and legumes was first discovered only a hundred years ago, farm-

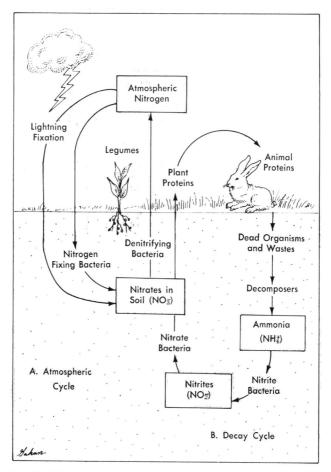

FIG. 9. *The nitrogen cycle. (A) Atmospheric nitrogen is cycled through nitrogen-fixing bacteria and lightning. (B) Nitrogen is cycled through decomposers and nitrite bacteria.*

ers have practiced rotation of crops for the past 2,000 years in order to obtain heavier crops of wheat. This rotation consisted of first growing legumes such as clover and then plowing them back into the soil. They had no idea why this practice improved their crops, but we know now that it is due to the nitrates made by the nitrogen-fixing bacteria. Many free-living microorganisms in soil can also fix nitrogen and make it available to higher plants. These include certain bacteria and blue-green algae that grow well in warm, shallow waters and have been found useful in fertilizing paddy fields to improve their capacity for growing rice.

A number of research projects currently are being conducted as part of the IBP program in the area of nitrogen fixation. In Venezuela, for example, studies will concentrate on the nitrogen-fixing systems of leguminous plants. In Puerto Rico, the role of epiphyllae in the cycles of a tropical ecosystem is being investigated. These leaf algae, lichens, fungi, and liverworts are being tested for ability to fix atmospheric nitrogen by the use of radioactive N^{15} and in other ways.

It has been estimated that nitrogen-fixing microorganisms fix between ½ and 3 kilograms of atmospheric nitrogen per acre per year; the figure may reach 100 kilograms in the most fertile regions. In addition to this figure, about 2 kilograms of nitrogen per acre per year is also fixed by lightning-discharge thunderstorms. The energy of a lightning discharge combines nitrogen with oxygen to form nitric oxides, which combine with water vapor to form nitrous and nitric acids. When these acids are brought down into the soil by rain, they are converted into nitrites and nitrates.

The atmospheric cycle is completed when denitrifying bacteria in the soil convert nitrates back to atmospheric nitrogen. An industrial process currently in use combines atmospheric nitrogen with hydrogen to form ammonia, which can then be changed into nitrates for fertilizer. This has increased the supply of soil nitrogen in agricultural ecosystems but interferes with the nitrogen cycle because the denitrifying bacteria are not able to keep up with this increased pace of restoring atmospheric nitrogen.

In the decay cycle, shown as B in Figure 9, grass absorbs nitrate ions from the soil to build a number of amino acids from which grass proteins are synthesized. When eaten by a rabbit, the proteins are digested into their amino acids, which are then recombined in new patterns to form proteins specific for rabbits. Similarly, when the rabbit is captured by a hawk, the process is repeated to build hawk proteins. The nitrogen is returned to the soil when bacteria and fungi of decay decompose the proteins and other complex compounds of dead plants and animals into ammonia gas (NH_3), which, in the presence of water, produces positively charged ammonium ions (NH_4^+). Animal feces and urine are other sources of nitrogen that are returned to the soil. Nitrite bacteria change the ammonium ions to nitrite ions, and these, in turn, are transformed by nitrate bacteria into nitrate ions. The cycle is now completed and nitrate ions are once again available to the grass. In addition, ammonium ions can also be absorbed and synthesized into proteins.

The Sulfur Cycle

Sulfur, the fifth element on the list of macronutrients, is an essential constituent of some of the amino acids and is, therefore, incorporated into proteins during their synthesis. Like nitrogen, sulfur is absorbed as a negative ion, the sulfate ion ($SO_4^=$). Unlike nitrogen, the cycling of sulfur does not involve the atmosphere at all but depends entirely on the activities of microorganisms. We will describe the sulfur cycle in an aquatic ecosystem (Figure 10). Sulfate ions are absorbed by the phytoplankton and incorporated into some of its proteins. As we go up the consumer food chain, the sulfur becomes part of fish proteins. When aquatic organisms die, decomposers break down the sulfur organic compounds into hydrogen sulfide (H_2S) in the bottom sediments. Special types of sulfur bacteria then oxidize the hydrogen sulfide back to sulfate ions, and the cycle is ready to begin again.

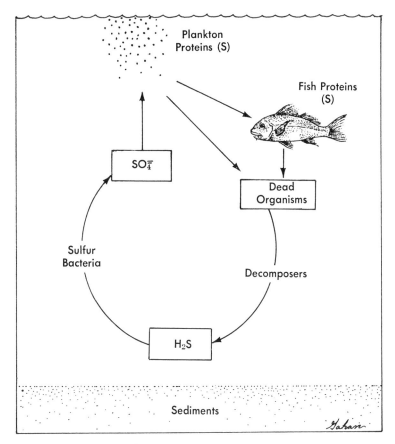

FIG. 10. *The sulfur cycle is an aquatic environment.*

The Phosphorus Cycle

The next element on the list of macronutrients, phosphorus, is a very important component of the complex nucleic acids DNA and RNA, whose molecules carry the genetic code responsible for developing an organism's inherited traits. In addition, it is also an important part of the ATP molecule, the energy carrier, and a constituent of phosphate salts found in bones and teeth. Phosphorus, like nitrogen and sulfur, is absorbed as a negative ion, the phosphate ion ($H_2PO_4^-$). The cycling of phosphorus illustrates what we might term a "half-cycle," because land ecosystems tend to lose it to the deep sediments of the sea. As Figure 11 shows, decomposers break down dead organic matter containing phosphorus into phosphate ions, which are absorbed by plants and synthesized into DNA, RNA, and ATP. These are then used by

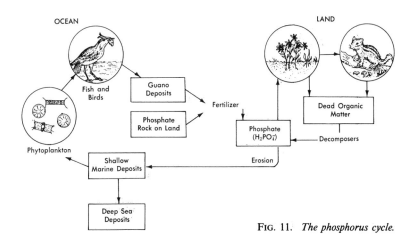

FIG. 11. *The phosphorus cycle.*

animals to make similar compounds and the phosphate salts of bones and teeth.

However, surface running water is responsible for eroding the soil and transporting a great deal of phosphate to the deeper sediments of the ocean. Deforestation and poor farming practices have contributed to the increased rate of such soil erosion. This loss is partly recovered on land by the weathering of phosphate rock, fallout from volcanic eruptions, and salt spray from wind-blown ocean waves. Upwelling helps to bring phosphates from the deeper sediments of the sea up to the surface waters, where phytoplankton assimilate them into organic compounds. These compounds feed fish, and, in turn, the fish-eating birds excrete guano on their nesting grounds off the west coast of South America. Large deposits of phosphate-rich guano have accumulated in some places to a depth of several hundred feet and are good sources of fertilizer for agricultural fields. Despite all this, there is still a net loss of phosphate from the land to the sea. We can compensate partly for this loss by mining phosphate fertilizer from phosphatic rock deposits. However, millions of tons of fertilizer produced this way each year are also washed into the ocean. Perhaps the solution to this problem will come from oceanic research conducted as part of the IBP program.

A group of IBP scientists are investigating the biological productivity of upwelling ecosystems, such as the area off the coast of Peru. Here, the anchovy fishing industry has grown in a few short years from one of negligible proportions into the largest single fishery in the world. Such sites also occur in the eastern and western parts of the nutrient-poor Mediterranean Sea, where the upwelling is thought to be responsible for fairly good fishing conditions. One of the important aims in this program is to assess the practicability of artificial upwelling. This could be a major contribution toward finding more food to feed man. In a recent study of an estuary ecosystem along the coast of the southeastern United States, it was discovered that ribbed mussels were important in the cycling of phosphorus. With each tide, as the mussels filter the seawater that covers the salt marsh for food, a large amount of detritus is removed from the water and deposited as sediment on the surface of the marsh. The particles are rich in phosphorus, other minerals, and vitamins and are important in the productivity of the ecosystem's plants.

The Cycling of Potassium, Calcium, and Magnesium

Potassium, calcium, and magnesium, the last three of our macronutrients, are absorbed as positive ions, K^+, Ca^{++}, and Mg^{++}. In land ecosystems, clay and humus particles are now considered to be important agents in cycling because their negative charges attract positively charged ions and keep them from being washed or leached out of the ecosystem. These ions become available to plants when living roots produce positively charged hydrogen ions (H^+) that move through the soil solution and displace them from the surfaces of the clay and humus particles.

Since potassium, calcium, and magnesium are all involved in the growth and development of a leaf, let us follow the cycling of these ions in a forest. After being absorbed from the soil water by the roots of an oak tree, they are conducted up the stem to a developing leaf, where the potassium controls cell division, the calcium becomes part of the middle lamella layer of plant cell walls, and the magnesium is assimilated into chlorophyll molecules. At the end of the growing season, the leaf drops to the forest floor. Here, decomposition returns the nutrients to the soil, where they are again available for absorption.

The Micronutrients

The micronutrients have major roles as essential parts of various enzyme systems. Chlorine seems to play a part in the production of ATP. Sodium seems to be needed by beet plants for the production of larger roots. If a plant does not have enough boron, its meristematic cells—responsible for the development of roots, stems, and leaves—die. Some micronutrients, not listed in Table 3, are needed for specific purposes in some plants. For example, the one-celled diatoms of phytoplankton need silicon to build their beautifully sculptured shells, and in terrestrial ecosystems, the same silicon helps to stiffen the stems of grasses. The micronutrients all occur as mineral salts, are absorbed into plants as ions, and follow cycles similar to sulfur and potassium.

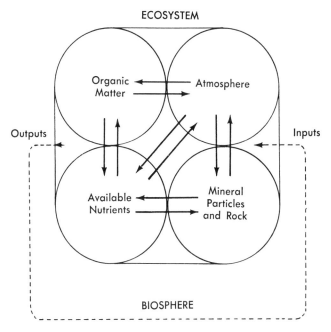

FIG. 12. *A nutrient cycling model.*

A Model for Nutrient Cycling

Modern ecologists find it easier to study the movements of nutrients within an ecosystem by referring to specific portions called compartments. For a terrestrial system, the model, as illustrated in Figure 12, locates the nutrients in four compartments: the organic compartment, the available nutrient compartment, the mineral particle and rock compartment, and the atmospheric compartment. In the organic compartment, the nutrients are grouped into macroscopic and microscopic organisms, both living and dead, and organic debris. The available nutrient compartment is composed of the ions of nutrients that are either in the soil solution or held at the surfaces of clay and humus particles. The mineral particle and rock compartment contains nutrients that are temporarily unavailable to living organisms because they are bound within the mineral particles and rock. In time, as a result of weathering, the surface nutrients are released as ions into the available nutrient compartment. The atmospheric compartment consists of gases that are found above and in the ground.

The arrows indicate the movement of nutrients from one compartment to another. When an ecologist investigates the cycling in an ecosystem, he measures how much of each nutrient is in each of the four compartments and how much of each nutrient is exchanged between compartments per unit time. The rates of movement of nutrients are more important in determining biological productivity than the amount present in any given place. The research scientist finds radioactive isotopes extremely helpful in following the rates of movement of tagged atoms as they are exchanged among compartments.

The diagram also shows the movement of nutrients from one ecosystem to another. When materials enter an ecosystem they are considered inputs, and when they leave they are called outputs. The agents that bring about this movement are either meteorological, geological, or biological. A leaf blown into a lake represents a meteorological input. Soil eroded by surface runoff illustrates a geological output. An animal that feeds on plants in a meadow and then drops fecal matter in a forest is involved in biological output from the field and biological input into the forest.

When constructing a nutrient budget for an ecosystem, the ecologist must therefore take into account how much of each nutrient enters and leaves the system per unit time. For example, let us assume that rain and snow are the main meteorological carriers of inputs and stream water is the major geological transporter of outputs. To determine if the ecosystem was gaining or losing a given nutrient, ecologists would measure the amounts of incoming rain and snow and outgoing stream water and then analyze them for nutrient concentrations.

IV

The Limits of the Environment

Chapters II and III focused attention on the role of great physical and chemical forces in the ecosystem. They described the importance of the flow of energy and the cycling of nutrients in limiting and regulating the community. We also learned that the community, in turn, interacts with the nonliving environment and thereby helps to control the rate of flow of energy and materials. For example, nitrogen-fixing bacteria can change atmospheric nitrogen into nitrates and make it available to plants.

A third aspect of community regulation involves the limits set by the physical environment. This was suggested in Chapter I when we pointed out that the nonliving environment of a forest, which included such factors as light, moisture, and temperature, affected the ability of an organism to live and reproduce there. The concept of limiting factors goes back to the time of Justus von Liebig, a German chemist, who in 1840 experimented with the effects of inorganic chemical fertilizers in agriculture. His "Law of the Minimum," originally intended to show the limiting effect on growth of factors present in the minimum quantity, has been extended to include the limiting effect of the maximum and the concept of interaction of factors.

The limiting factor concept helps the ecologist approach the study of complex environmental problems. For example, oxygen becomes a limiting factor only in certain environments. Therefore, if fish are dying in a polluted lake, the concentration of oxygen in the water would be one of the first things to investigate because the dissolved oxygen supply in water is easily used up and is very often in short supply. On the other hand, if mice are dying in a meadow, the ecologist would investigate some other cause because atmospheric oxygen is quite constant and not likely to be limited.

Every species can live within a range of values for each factor in its environment but cannot survive outside the low and high values called the limits of tolerance. For example, the limits of tolerance of frog eggs for temperature are 0° C and 30° C. A species will stay alive as long as all the factors in its environment are within the limits of tolerance for that species. However, for most species, there is a best value for each factor called the optimum. For example, the optimum temperature for frog eggs is 22° C, because more eggs hatch at that temperature than at any other.

In some organisms, when one factor changes from an optimum to a lower concentration, tolerance to another factor may also be reduced. For example, during the cold winter of 1963, certain marine snails in the Blackwater Estuary of Ireland died because at the low temperatures they became less tolerant to the low salinity of estuary waters. In a similar way, a deficiency of nitrogen has been found to reduce the tolerance of grass to drought.

Investigating Limiting Factors

Because of the interaction of factors at the community level, an ecologist must exercise caution in transferring to the field information on limits of tolerance for a factor carefully determined in the laboratory. It would be necessary to study the situation in the field and then go back to the laboratory for experiments with other factors and combinations of factors.

A novel way of studying limiting factors in the open sea was introduced by investigators from the Woods Hole Oceanographic Institution in connection with the study of primary production in the phytoplankton of the Sargasso Sea, a desert area of the ocean lacking in nutrients. They set up enrichment experiments on board a research ship and added nutrients one by one to water containing the natural phytoplankton. They discovered that food production, measured by the uptake of carbon-14, did not increase with the addition of nitrates and phosphates unless the water was enriched first with the micronutrients silicon and iron.

Limiting factors in agricultural research currently are studied in two ways. In one, sophisticated collections of greenhouses called phytotrons simulate a great variety of outdoor conditions. In the other approach, outdoor experimental fields, like the one at Rothamsted, England, provide test plots for crops in which the minerals are carefully controlled. The U.S. National Committee for the IBP, heavily involved in scientific efforts to increase agricultural production, is considering ways of increasing the area now being cultivated—only 7.6 percent of the earth's land. The scientists are investigating possible means of dealing with the four major factors that limit agricultural production—drought, root development, low temperature stress, and lack of heat tolerance. For example, in the single most limiting factor to plant distribution, low temperatures, areas of research are suggested such as genetic development of crop plants resistant to low temperature and basic physiological research to find out how chilling or freezing kills plants and how some plants acclimate to resist these stresses.

Ecological Indicators

A study of the species in a given ecosystem can often indicate whether or not unfavorable environmental conditions are developing. In many instances, rare species that have narrow limits of tolerance for a particular factor make the best indicators. For example, range managers have discovered that a decline of certain rare species of plants with a low tolerance for grazing indicates the beginning of overgrazing before it is apparent in the grassland. The best indicator of conditions, however, is the species diversity of an entire community.

Stable, healthy communities are characterized by the presence of a few species with many individuals and many species with a few individuals. Environmental stress will affect the distribution of animal species in such a way that it can be summarized as follows for water pollution:

1. Great variety with a few of each kind = clean water
2. Less variety with great abundance = moderate organic pollution
3. One or two kinds only, with very great abundance = severe organic pollution

Investigators studying water pollution in a salt marsh, for example, will seek to determine if there is a decline in the number of species or in a diversity index because this often indicates pollution before the total number of individuals are severely affected. The diversity index is obtained by determining the total number of different species (cumulative taxa) in samples collected and dividing it by the square root of the total number of individuals in the samplings. If d represents the diversity index, then

$$d = \sqrt{\frac{\text{cumulative taxa}}{\text{total individuals}}}\,.$$

For example, Table 4 shows the data obtained by a group of students who compared the species diversity in two salt marsh ecosystems. They concluded that Orchard Beach was more polluted than Sunken Meadow because its diversity index was smaller.

DATA OBTAINED IN MEASURING THE DIVERSITY INDEX OF
TWO DIFFERENT SALT MARSHES

Measurement	Sunken Meadow	Orchard Beach
1. Cumulative Taxa	20	10
2. Total Number of Individuals	218	80
3. Diversity Index	1.354	1.118

TABLE 4

Light as a Limiting and Regulatory Factor

The amount of available sunlight is a limiting factor for plants. In ocean waters, phytoplankton are limited to the surface layer because light is absorbed as it penetrates through water. Light is also an important regulator of the daily and seasonal activities for both plants and animals. The length of day, or photoperiod, acts as the trigger that sets off the biological clock by which organisms time their activities in temperate zones. This is possible because, for a given latitude and season, the photoperiod is always the same.

An example of plant activities controlled by photoperiodism is flowering. Plants like chrysanthemums and asters are called short-day plants because they bloom and set fruit in the autumn, when the days are short and the nights long. Spinach and barley, on the other hand, are called long-day plants because they start to flower when exposed for two weeks to days that are 13 to 14 hours or more long but will not flower if the days are less than 13 hours long. If the darkness period of a short-day plant is interrupted with a brief exposure to light, it will not flower. It has been discovered that the day length acts through a pigment called phytochrome, which responds to red light and, in turn, activates enzymes controlling flowering processes.

Many animals are also affected by the relative lengths of day and night. For example, the photoperiod has been shown to be the timer that sets off such activities as molting, fat deposition, migrations, and breeding in certain migratory birds whose breeding and wintering areas both lie in the North Temperate Zone. An example is the white-crowned sparrow, which has been studied for many years. The approach of the long days of late winter or spring stimulates in this bird the activities mentioned. Biologists have been able to produce out-of-season migratory restlessness, fat deposition, and a maturing of the sex organs in midwinter in the laboratory by an artificial increase in the light period.

The photoperiodic mechanism in birds is believed to act through the hypothalamic part of the brain by way of photoreceptors in the eyes and diencephalon. Here a neurohormone is produced that stimulates the pituitary gland to release gonadotropic hormones, which are carried by the blood to the gonads. This is illustrated in Figure 13. Other seasonal activities that are under photoperiod control are hibernation and seasonal change in mammalian hair coats.

Temperature as a Limiting and Regulatory Factor

Most living things are active between 0° C and 40° C; for many species the range is even smaller. The slowing down of chemical reactions in living things as the temperature drops explains their inactivity at lower temperatures. This is why plant productivity drops during the winter in temperate zones and some animals hibernate. Although chemical reactions proceed faster at higher temperatures, they slow down and stop above 40° C in living things because the enzymes that catalyze the reactions are destroyed. In spite of all this, some forms of life are able to survive outside the 0°–40° C range. For example, dormant stages such as the seeds of plants and the spores of some bacteria can withstand temperatures as low as −75° C, whereas certain algae live in the

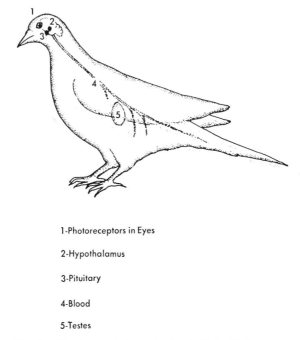

1-Photoreceptors in Eyes

2-Hypothalamus

3-Pituitary

4-Blood

5-Testes

FIG. 13. *The photoperiodic mechanism in birds. (1) Photorecep-*
tors in eyes stimulate the hypothalamus. (2) The hypothalamus
secretes a neurohormone that stimulates the pituitary gland. (3)
The pituitary gland secretes a gonadotropic hormone into the blood
(4), which transports it to the testes. (5) The testes of the male bird
mature.

near boiling temperatures of about 92° C in hot springs and gey-
sers. What decides whether or not a given species can live in a
particular location is often the fluctuations in temperature that
occur in that region from season to season, within a season, and
even within a single day. For example, the eggs of brook trout can
develop only in water whose temperature is between 0° C and 12
° C, whereas the eggs of frogs can develop in water whose temperature goes as high as 30° C.

Cold blooded animals have body temperatures that are very
close to the temperature of their surroundings. As a result, they

become inactive in cold weather and in very hot weather will move into the shade or hide in a hole. Cold blooded animals include all the invertebrates as well as the fish, amphibia, and reptile classes of the vertebrates. Warm blooded animals, represented by birds and mammals, can maintain a constant body temperature in all seasons and therefore remain active at all times. This gives them two advantages. First, they are less vulnerable to attack by predators than are cold blooded animals, which are subject to periods of inactivity. Second, they have successfully colonized places on earth with very different climates and are therefore less likely to become extinct if the climate suddenly changes in one of these regions.

Like light, temperature may also be a factor in regulating the seasonal activities of organisms. The seeds of winter cereals, sown in autumn, need the cold temperatures of winter before they can flower the next summer; failure results if the seeds are planted in the spring. In like manner, the bulbs of tulips, daffodils, and hyacinths also need a period of cold before they can begin to grow and produce the next season's flowers.

Rainfall as a Limiting and Regulatory Factor

Rainfall is important to the survival of terrestrial plants and animals because it is the ultimate source of water, without which the chemical reactions of life could not go on. The various adaptations of desert plants and animals will be described in Chapter VII. In some parts of the tropics and subtropics where there are wet and dry seasons, rainfall is the main factor regulating the seasonal behavior of living things, especially their reproductive cycles. In contrast, the roles of light and temperature in governing seasonal behavior are characteristic of temperate climates. Another example of regulation by rainfall involves the seeds of many annual desert plants, such as cheat grass, that germinate only when a shower produces one half inch of rain or more. The rainfall washes out of the seed coat a chemical substance that inhibits the sprouting of the seed.

The flora and fauna (plants and animals) that can live in a given environment are often determined by the ratio of rainfall to evapotranspiration rather than by rainfall alone. Two areas of equal rainfall may have various kinds of plants because of unequal rates of evapotranspiration. Thus, on the basis of the ratio, an area may tend to be a grassland, a desert, or a forest.

Fire as a Limiting and Regulatory Factor

In warm or dry regions and in regions with warm and dry seasons, periodic small fires favor the survival and growth of one species over another. In the southwestern United States, for example, fire favors grass over mesquite shrubs, which otherwise would crowd out the grass. This is illustrated in Figure 14. Another

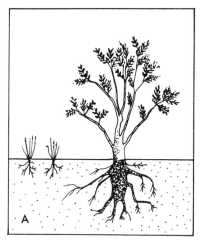 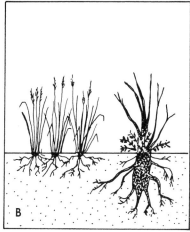

FIG. 14. *How fire favors one species over another. (A) In the absence of fire, the mesquite shrub grows faster than the grass and tends to crowd it out. (B) After a fire, the grass recovers quickly and produces a flush of new foliage. Because fire favors the grass over the mesquite, the grass is called a "fire type."*

example of a "fire type" species is the long-leafed pine of the southeastern United States, a valuable timber tree that tends to be crowded out by scrub hardwoods in the absence of fire. But periodic fires destroy the competitors and enable the pine to take over because the delicate tips of its growing shoots are encircled by fire resistant pine needles.

In hot or dry regions, fire can also act as a decomposer to help release minerals from surface debris in grassland or forest that has become so dry that bacteria and fungi are unable to break down the litter. The big game herds of Africa and the deer in the California chaparral thrive best when periodic fires release the nutrients in the dry litter and help produce a flush of new grass or foliage. These periodic fires keep the surface debris to a minimum and thereby also prevent the start of severe fires that might spread to destroy homes.

V

The Interactions of Populations

In the IBP study of the grassland ecosystem at Pawnee, researchers found that Pawnee's most prevalent bird, the lark bunting, likes to build its nest on the eastern side of a shrub called the saltbush. Should the saltbush disappear from Pawnee, so would the lark buntings. And if this happened, what would eat the insects and keep their numbers down? The IBP researchers are trying to understand such interactions so they can predict and ease the effects of man-made changes on the environment.

Population Counts

Although interested in individual plants and animals, ecologists find it more useful to study whole populations to determine how biological factors affect their size. To study changes in a given population, they must first make measurements to determine its density. Density is the number of individuals of the same species found in a given amount of space. In some cases, it is possible to count the total number of individuals inhabiting a given area. For example, the number of elephants occupying 100 square miles of open plains in Africa can be directly counted from a helicopter or a light plane. When direct counts are too tedious or impossible, population density can be estimated by the use of a sampling scheme.

In one such scheme, total counts are determined by using limited sample areas called quadrats. If we wish to determine the number of earthworms in the ground, for example, we would mark off a square meter and then count the worms, first in the litter on top of the soil and then in a layer of soil removed from this area. We would then repeat this procedure in several areas to get an average number of worms per square meter per given depth.

A sampling scheme commonly used for small animals such as mice and chipmunks is the mark and recapture method. In this technique, a sample of animals, such as a hundred field mice, is captured from a given area. The mice are marked with metal tags or dye. They are then released back into the population and given time to mix with unmarked individuals. Another sample of a hundred field mice is then trapped and the number of marked individuals determined. Let us assume that ten marked mice are found among the second hundred. An estimate of the total population can then be calculated from the following proportion:

$$\frac{\text{Total Population Number}}{\text{Number Marked and Released}} = \frac{\text{Number Recaptured}}{\text{Number Marked in Recapture Sample}}$$

In this instance, the total population would be $\frac{100}{10} \times 100$, or 1,000 field mice. This sampling method, however, is based on the assumption that the released individuals spread out randomly through the population and are not subject to different recapture rates.

When population density is being measured, ecologists also record the dispersion, or the way individuals are distributed in space. Three patterns of dispersion are measured: clumped, uniform, and random. The most common type is the clumped pattern, in which individuals occur in groups. Examples would be plants that spread by runners and animals that come together to share a common source of food, water, or shelter. The characteristic wide spacing between desert plants, which results from competition for water, illustrates the uniform pattern of distribution in which individuals are evenly spaced. Plants that have little effect on one another and do not spread by runners exhibit the random type of dispersal in which an individual is just as likely to be in one location as in any other.

Population Change

To determine if a population is increasing or decreasing, we must know the relative rates at which individuals arrive through birth or immigration and are lost through death or emigration. The birth rate, or natality, refers to the rate at which individuals are added to a population by reproduction. The death rate, or mortality, refers to the rate at which individuals are lost from a population by death. To determine the birth rate, you divide the number of births in a given year by the total population at the midpoint of the year and multiply by 1,000. Similarly, the death rate is expressed as the number of deaths per 1,000 individuals in the population.

A comparison of the birth and death rates should indicate if a population is increasing or decreasing. But to understand the change more completely, we must know the ages of the individuals that are dying. With respect to age distribution, it is often noted that in many animals mortality is highest among the very young and elderly and lowest among the middle age group. For example, studies of a black-tailed deer population living in a chaparral biome reveals that of each 1,000 deer born, about 370 die during the first year, half survive into their fourth year, about one third survive into their seventh year, and one fifth survive into their ninth year. If a deer reaches its ninth year, it has a life expectancy of about one and a half years. This pattern of mortality can be represented graphically as a survivorship curve by plotting the number of deer surviving at each age on a graph.

Population Growth

Now let us see how changes in population density take place in nature. If food and shelter are plentiful, birth rates will be high and death rates low. Some species have the biotic potential to reproduce very rapidly and produce enormous numbers of offspring. For example, meadow mice reach sexual maturity at thirty days and can produce as many as seventeen litters of six to nine young during the warm months of a year. Under ideal conditions of abundant food, water, space, and protection from predators, they could increase their population to crowd the entire earth within a dozen generations. This will never happen, however, because as a population increases, the death rate begins to rise. More and more individuals will die of starvation, lack of shelter, predation, disease, and overcrowding. In addition, the health of the females of the species will be affected and the birth rate will begin to fall.

If we plot a typical population's density along the Y axis against time along the X axis, a graph will reveal an S-shaped growth curve, as shown in Figure 15. This curve shows that a population's growth is slow at first, becomes rapid as the species approaches its biotic potential, and finally levels off as further growth is limited by the environment. When the birth and death rates become equal, the population is said to be in equilibrium with its environment.

Population Irruptions

The simple S-shaped growth curve, typical for most populations of plants and animals, can also be observed in laboratory experi-

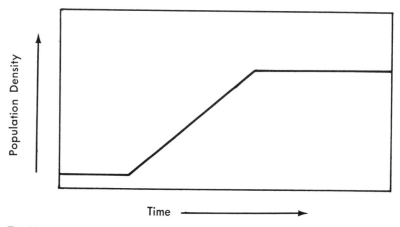

FIG. 15. *A typical S-shaped population growth curve. Growth is slow at first, then more rapid as the biotic potential is approached, and finally levels off as the environment limits further growth.*

ments with yeast cells in which a small population is placed in a culture medium with an ample supply of food. But for some species in nature, the curve often fluctuates instead of leveling out. This happens when a population, under very favorable conditions, increases so rapidly that its density exceeds the carrying capacity, or maximum number of individuals, that can be supported by an environment. Such a rapid rise in population is called an irruption. As the supply of food dwindles and shelter becomes scarce, deaths exceed births and the population declines. When conditions again become favorable, the sequence is repeated. In most cases, the fluctuations will continue to get smaller and eventually approach a stable level.

If the irruption produces a very severe crisis, the population may fall to such a low level that it will recover very slowly or not at all. A minimum population level is necessary for successful reproduction. If the density becomes so low that members of the opposite sex fail to meet each other, reproduction will not take place and the population will die out. In some species of sea gulls, the behavior patterns needed for pair formation and care of the young are stimulated by the presence of other members of the population. Oysters require a fairly large population for successful propagation for another reason. They start life as floating larvae that must settle on the shells of the older oysters in the colony or they will not survive. The lesson for the conservationist who is interested in perpetuating oysters, and similar species that show low population growth rate at low density, is to make every effort to keep a large standing crop at all times.

There are a number of animals in northern climates that experience population irruptions and crashes in cycles. The most famous one is the small, mouse-like rodent of the Canadian arctic, the lemming, whose population shows a sharp increase every three or four years. Others are the snowshoe hare and its predator, the Canadian lynx, both well known for their population fluctuations, which are repeated in cycles of nine to ten years. Estimates of their populations are based on records of the skins of these animals sold by trappers to the Hudson Bay Company. A graph that is plotted from these estimates, which date back to 1845, is shown in Figure 16. Note that a rise and fall in the hare population is always followed by a corresponding rise and fall in the lynx population. This is easily explained by the predator-prey relationship between the two animals. But the reason for the hare fluctuations is more difficult to comprehend. Ruffed grouse, pheasants, quail, and muskrats seem to go through similar ten-year cycles. Although the explanation is still not clear, most ecologists believe that a closer study of population interactions will reveal the answer.

Intraspecific Competition

In spite of the fluctuations in some species, the size of most populations tends to be close to the ecosystem's carrying capacity and remains remarkably constant year after year. The biological

Population

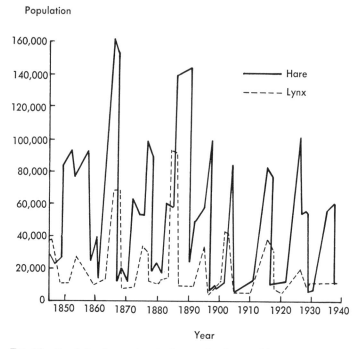

Year

FIG. 16. *Population fluctuations in the snowshoe hare and lynx. A rise and fall in the hare population is always followed by a corresponding rise and fall in the lynx population.*

factors responsible for this regulation will now be considered.

Members of the same population will strive for a resource that is not in adequate supply. For example, in a forest, plants strive for light and nutrients and animals compete for food and shelter when they become scarce. This type of competition among members of the same species is called intraspecific competition, and it tends to limit population size. Before severe competition develops for food and space, however, overcrowding may become a factor in reducing the population. Rats reared in crowded pens fail to develop normal behavior patterns of courtship and mating, and the disturbed females build unfinished nests in which the young are neglected. High infant death rates and lowered fertility both act to keep the population density low even though plenty of food, water, and nesting material is available.

Overcrowding, which might lead to severe competition for water, is prevented in desert plants by the secretion of inhibiting chemicals into the soil. These chemicals kill seedlings that have begun to sprout and enable shrubs to be spaced apart.

Another form of intraspecific competition in many animals is the phenomenon of territoriality, which represents a behavior pattern in which the male stakes out an area at the beginning of the breeding season and drives other males of the same species out of the territory. This procedure is a prerequisite for successful mating and rearing of the young and restricts reproduction to the stronger and more dominant animals. The weaker individuals either do not reproduce or are crowded into areas where the chances of survival of their eggs and young are poor. In most cases, warning signals are sufficient to keep intruders out, and little fighting is actually involved in the defense of the territory. Birds sing loudly or show their brightest feathers as a threat display. Coyotes and wolves urinate at different points along the edges of their territories. Rabbits establish territoriality with their droppings. An invader respects an established territory and readily withdraws when threatened.

Interspecific Competition

In Chapter I we briefly discussed the concept that when two different species compete for the same food and space, in most cases one species will be successful in eliminating the other. Deer can keep moose from increasing by competing with these larger animals for browse. Two different species may also compete for nesting places. Raccoons, squirrels, and even wasps compete with wood ducks for nesting cavities.

The most spectacular form of interspecific competition involves predation in which one organism, the predator, kills an animal of another species, the prey, and consumes all or part of it. Predators are actually beneficial to the prey species because they weed out the weak and sick members. They also prevent grazing and browsing animals from becoming so numerous that they destroy their own food supplies and die of starvation.

Most ecologists consider changes in predator populations to be an effect of comparable changes in prey populations rather than a cause of the fluctuations. Look again at Figure 16, which compares fluctuations in the populations of snowshoe hares and lynxes, and observe that each rise and fall in the hare population precedes a comparable rise and fall in the lynx population. In a stable ecosystem, where a balance has been established between a predator and a prey population, an increase in prey numbers will cause an increase in the rate at which prey are captured by the predators. Once the increase in the prey population has been slowed, the rate of attack by predators will decrease.

Some species of deer seem to be strongly regulated in this way by predators. When the natural predators—such as wolves, bobcats, and pumas—are removed, the stability of the ecosystem is disrupted and the deer multiply so rapidly that they eventually overgraze the forest and many die of starvation during the winter. In wild regions, predators should be protected not only for the good of the deer population but also for the good of the forest.

Parasitism is similar to predation in that both can be factors in regulating populations. However, there are a number of differences between them. A parasite is a small organism that lives in or on its host, which serves as both a source of food and a habitat. In contrast, a predator is free living and larger than its prey, which serves as a food source but not as a habitat. Predators are carnivores, but parasites may be animals or plants. Many parasites—like hookworms, tapeworms, and fleas—are invertebrates, but some—like the lamprey eel—are vertebrates. Most plant parasites are fungi, but some—like mistletoe and dodder—are flowering plants. The smallest parasites are microorganisms belonging to such groups as the viruses, bacteria, and protozoa. Many parasites can live only in one or a few related species and can quickly adjust to increases or decreases in host numbers. On the other hand, predators are more generalized. A wolf, for example, may choose its prey from a varied bill of fare that includes deer, moose, rabbits, rats, squirrels, and birds. Predators are important as regulators in the ecosystem as a whole, but parasites are more effective in limiting a specific population. Insectivorous birds are important in controlling all the insect populations in a forest. But if a particular species irrupted, the parasite specific for this host would be far more effective in inhibiting the population.

Positive Interactions

There are many interactions that involve cooperation rather than competition, and these may be termed positive interactions as compared with the negative interactions we have been considering so far in this chapter. Both types of interactions are equally important, and a balance between them in populations is needed to achieve stability in a well-regulated ecosystem.

In one type of positive interaction called commensalism, one organism benefits from an association with another organism that is neither helped nor harmed. For example, the remora fish attaches itself to a shark by means of a dorsal suction disc and is transported to the scene of a kill, where it feeds on scraps left by the shark. The remora benefits but the shark is unaffected. Similarly, barnacles attached to the skin of a whale are carried to areas where more food may be available. In another type of commensal association, a number of marine organisms find shelter in the cavities and canals of sponges, but do neither harm nor good to the host.

A second type of positive interaction called protocooperation involves a casual association in which two populations help each other but do not need each other for survival. Sea anemones, for example, often attach themselves to the backs of crabs, where they provide camouflage and protection for the crabs. In return, the sea anemone obtains scraps of food when the crabs capture and eat other animals. In another example, several kinds of African birds perch themselves on the backs of rhinoceroses, elephants, and other large mammals and feed on the lice and ticks in the mammal's skin. In return, the mammals are relieved of their parasites and may even get the benefit of danger warnings if the birds suddenly take off in fright.

In the most advanced type of positive interaction, called mutualism, two populations not only benefit each other but also need each other for survival. One example, the mutualistic association of nitrogen-fixing bacteria and legumes, has already been described in Chapter III. Another example of mutualism is the close relationship between wood-eating termites and their intestinal flagellates. Without these protozoa, the termites would starve to death because the flagellates secrete an enzyme, lacking in termites, that can digest cellulose. The termites depend on the flagellates for digestion and the flagellates need the termites for their wood supply.

Lichens, simple plants that can live on bare rock outcrops and the frozen soil of the arctic tundra, are close associations of algae and fungi. The fungi provide shelter, water, and minerals for the green algae, and, in return, the algae produce food for the fungi. There is evidence in the study of different lichens that mutualism evolved from parasitism in which the parasite host association became beneficial. In some primitive lichens, the fungal hyphae actually penetrate into the algal cells and behave as parasites of the algae. In more advanced species, the fungal hyphae surround

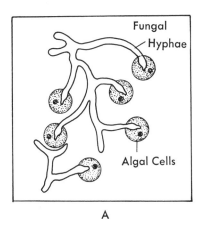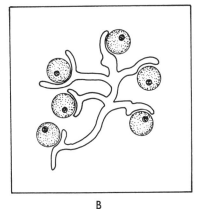

A   B

FIG. 17. *Evolution of mutualism in lichens. (A) The fungal hyphae actually penetrate the algal cells in some primitive lichens, a type of parasitism. (B) In more advanced lichens, the hyphae touch the algal cells but do not penetrate them, a true case of mutualism.*

the algal cells but do not break into them. This is illustrated in Figure 17.

Because of the possible evolutionary ties among parasitism, commensalism, and mutualism, many ecologists group all such close relationships under the heading of symbiosis. It is clear that all of them are important in establishing a stable equilibrium in an ecosystem.

VI

Ecological Succession

Ecologists have learned that communities change in composition as they interact with the physical environment and that gradually, over a period of time, they develop more or less permanent ecosystems. The principles involved in such community changes can best be understood by carefully observing the changes that take place in an abandoned farm field, such as might be found in the southeastern United States, as the field develops into an oak-hickory forest.

The first plants to grow on the abandoned field are the same weeds that competed with the farmer's crops of cotton and corn. These pioneers, many of them inadvertently introduced from Europe, Asia, and Africa, grow rapidly in clearings such as vacant lots, roadsides, railroad embankments, and abandoned fields. They grow quickly in sunlight, reach maturity in a few weeks, and produce thousands of seeds which are dispersed by wind, running water, and animals. These seeds are very hardy and can lie dormant on the ground for years until conditions are right for sprouting.

Crabgrass gets the jump on the other weeds and covers the ground within the late summer months following abandonment. It then begins to change the physical environment by adding shade to the soil and conserving moisture by reducing surface evaporation. These new conditions favor the sprouting of horseweed, which begins to grow in the moist shadows of the crabgrass. By the next spring, the horseweed is crowding out the crabgrass, and a third weed, wild aster, begins to appear among the first two. Although hardly noticed this first summer, wild aster flowers are producing large numbers of seeds.

By the second spring, the soil has been enriched by the remains of crabgrass and horseweed plants and conditions favor the growth of asters, which then dominate the field during the summer. Grasshopper sparrows and meadowlarks flock in to feed on the abundant seeds and remain to nest. A few mice also move in.

During the next spring, broomsedge grass begins to appear among the asters, and its tall clumps dominate the field during the third summer. Crabgrass is completely shaded out by the three-feet-high grass cover, but the asters remain, scattered between the clumps. Cotton rats move into the grassland to feed on broomsedge seeds, and deer may visit to browse. Figure 18 illustrates the four pioneers that help to establish a grassland in an abandoned field.

As the soil becomes richer in organic matter and denser in shade, shrubs begin to appear among the broomsedge and asters. They slowly increase in size and number, shade out the asters, and dominate the field by the fifteenth summer. Meadowlarks still nest in some remaining grassy spots, but field sparrows, yellow-breasted chats, and yellowthroats are nesting in large numbers in the shrubs. White-footed mice live and eat in the shrubs, and predators such as foxes and weasels eventually move into the area.

The grassland is almost gone by the twentieth year. The rich soil, shaded by the shrubs, now favors the sprouting of pine tree seeds, and the seedlings slowly push up in different spots. Although the young pine trees grow slowly, they outgrow the shrubs in a few years and begin to shade them out.

The shrubs are almost gone by the thirty-fifth year, and a pine forest stands in their stead. Pine warblers and Carolina wrens nest

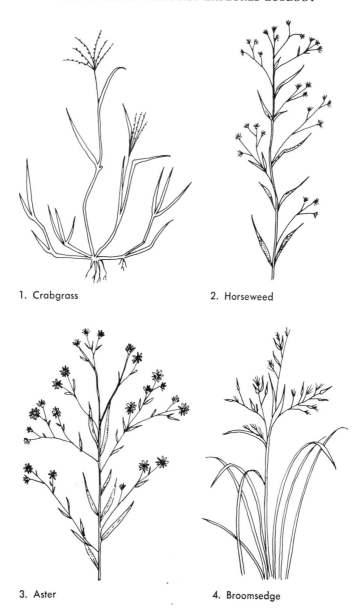

1. Crabgrass

2. Horseweed

3. Aster

4. Broomsedge

FIG. 18. *The pioneer weeds in the succession of an abandoned field.*

in the trees and are part of a forest community that also includes mice, squirrels, skunks, a variety of insects, and many other animals. The physical conditions of the pine forest floor are now favorable for the growth of oak and hickory tree seedlings, and they slowly grow under the pines.

By the hundredth year, in the climate of the southeastern United States, the oak and hickory trees have grown past the pine trees and are beginning to shade them out. In another fifty years, the pines are replaced by an oak-hickory forest similar to the one that was cleared away several centuries before to create the cultivated field. Its populations of plants and animals are very much like those of the undisturbed oak-hickory forests in the region which have demonstrated stability for hundreds of years. Figure 19 illustrates the ecological succession on an abandoned field.

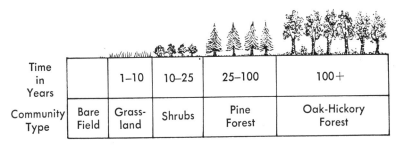

Time in Years		1–10	10–25	25–100	100+
Community Type	Bare Field	Grass-land	Shrubs	Pine Forest	Oak-Hickory Forest

FIG. 19. *Succession on an abandoned field in the Southeastern United States.*

Basic Ideas in Succession

Ecological succession may now be defined in terms of three ideas. First, it is a series of orderly changes that are predictable. For example, if we observe an area covered with shrubs in North Carolina, we can predict that it will be replaced by a pine forest but not by a grassland. Second, the succession results from the changing of the physical environment by the community. Each set of organisms modifies the local physical environment in such a way that conditions become more favorable for another set of organisms. Thus, in the example of the abandoned field, the grassland changed the conditions of the soil and made it more favorable for the growth of shrubs. Third, the succession ends in the establishment of a climax community that is most stable for the area.

The entire series of stages from pioneer to climax is called a sere. The species involved in the seral stages, the time required for the succession, and the degree of stability of the climax depend on the climate, soil conditions, topography, and other physical factors. In a temperate region with sufficient rain, the climax is usually a beech-maple forest. In the somewhat drier climate of the southeastern United States, the oak-hickory forest is the climax. In southwestern Michigan, the same climate will support beech-maple forests in moist soils and oak-hickory forests on coarser, drier soils. In this case, we call beech-maple the climax and oak-hickory a preclimax stage. Where a temperate climatic area, such as New York State, borders on a cooler, wetter climate to the north, the climax beech-maple gradually yields to hemlock trees to produce a postclimax stage.

When a seral stage persists because it is prevented from developing to a climax, it is called a subclimax. For example, the long-leaf pines of the southern Atlantic and Gulf coastal states are favored by periodic fires over the hardwood forests, which should be the climax. A grassland becomes a subclimax when men or rabbits destroy the tree seedlings that were destined to become a forest.

Secondary Succession

When we described succession on an abandoned farm, we indicated that cotton and corn were grown on the site for probably hundreds of years and that an oak-hickory forest was originally removed to create the farm. When succession occurs on sites previously occupied by organisms or on sites where nutrients and conditions of existence are already favorable, it is termed a secondary succession. Examples of such sites are abandoned croplands, plowed grasslands, burnt down forests, and ponds or lakes.

In a pond or lake area, as the lake fills in with silt from the surrounding land and organic debris from the organisms in the

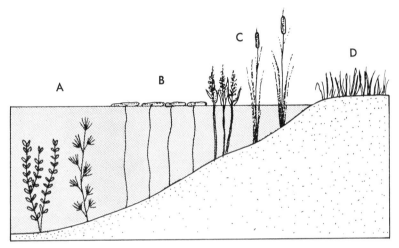

FIG. 20. *Succession in a lake. (A) Submerged plant stage (elodea, cabomba). (B) Floating leafed stage (water lilies). (C) Emergent plant stage (pickerel weed, cattails). (D) First terrestrial stage (sedge meadow).*

water, algae will be succeeded by larger, submerged water plants, like elodea and cabomba, then by floating-leafed plants, such as water lilies, followed by emergent plants, such as cattail and pickerel weed. The pond is now a marsh. Sedges and grasses take over as the marsh becomes a prairie or grassland. See Figure 20. From this point on, if the climate is that of the southeastern United States, the seral stages will include shrubs, pines, and a climax oak-hickory forest.

Primary Succession

When a succession occurs on barren ground where the conditions of existence are not at first favorable—for example, on a newly exposed sand dune or on a recent lava flow—it is called a primary succession. This type of succession has been extensively studied on the sand dunes at the south end of Lake Michigan. These dunes are the result of water currents that erode the western shore of the lake and deposit the sand on the beach at the southern end. When the sand dries during the summer, winds blow it inward to build up the sand dunes.

The environment of a sand dune is very hostile. A plant would have to contend with wind-shifted sand, very high surface temperatures, little available moisture (rainwater filters rapidly through the sand), and a poor supply of nutrients. In spite of these harsh conditions, some grasses are able to root in the sand and capture the dunes. Burrowing animals such as tiger beetle larvae and spiders, which are able to survive the high surface temperatures, also make the dunes their home. This community represents the pioneer stage in the succession.

As organic debris from these organisms builds up humus in the soil, more nutrients and moisture become available and conditions are made more favorable for seedlings of sand cherries, willows, and cottonwoods. In time, the cottonwoods become the dominant trees in the site. Many kinds of insects and birds find a home here in this seral stage. As the soil improves still more, pine seedlings begin to grow in the shade of the cottonwoods. After many years, the pines become taller than the cottonwoods and shade them out of existence. The seral stages are now very similar to the corresponding ones in the abandoned farmland as oaks and hickories gradually replace the pines. However, the climatic conditions at the southern end of Lake Michigan favor the ultimate development of a beech-maple forest.

Time Factor in Successions

Recent studies on primary successions indicate that the development of a climax will take at least 1,000 years, whereas a secondary succession will take about 200 years. When the climate is more severe, as in a tundra, desert, or grassland region, the sere may be of short duration because it is difficult for the community to change the harsh physical environment to any great extent. For example, it is possible for a person to witness during his lifetime the complete sere of a secondary succession in a grassland. The breakdown of the seral stages has been described as follows: (1) annual weed stage—two to five years; (2) short-lived grass stage— three to ten years; (3) early perennial grass stage—ten to twenty years; (4) climax grass stage—twenty to forty years. In ocean ecosystems, the community is unable to modify the physical environment to any great extent, and as a result succession is very brief, lasting only a few weeks. For example, in a marine bay, there is a brief succession from diatoms to dinoflagellates each season and sometimes several times a season.

Important Trends in Succession

Ecologists have constructed a model that depicts the trends in community changes as succession proceeds from pioneer stage to climax. One trend indicates a continuous change in the kinds of plants and animals found in different stages. For example, weeds are important in the pioneer stage on an abandoned farm, but they are not important in the oak-hickory climax. Some birds fill niches in only one seral stage. Thus, the grasshopper sparrow is found only in the grassland stage, and the yellowthroat occurs only in the shrub stage. Some birds, like the cardinal, have wider niche preferences and will persist through several stages. In general, however, the more species there are available for filling particular niches, the more restricted will be the presence of each species in different seral stages. This limitation is a result of the interspecific competition discussed in the previous chapter.

A second important trend in succession is the increase in the biomass of the community and in the amount of organic matter it produces. In a terrestrial ecosystem, this decreases the susceptibility to certain disturbances. For example, a steep hillside is less likely to be eroded if it is covered with vegetation that holds the soil in place. The organic matter also includes a vast number of soluble substances such as sugars, amino acids, and other compounds called extrametabolites. In the sea, some extrametabolites are the products of decomposers, others result from bacterial synthesis, and still others are derived from the excretions and secretions of living plants and animals. In the forest, many metabolites extrude from the roots of forest trees. Some of these substances may provide food for zooplankton or soil arthropods such as mites and pillbugs. Other substances, such as vitamins, may act as growth promoters, and still others, such as antibiotics, may act as inhibitors. A number of marine biologists have even suggested that these growth-promoting and growth-inhibiting substances may be the factors controlling the succession of organisms in the planktonic environment of the sea.

A third trend in succession is the tendency of species to increase in diversity. The increase in the diversity of heterotrophs is especially striking when we compare the huge variety of heterotrophs found in the later stages of succession with those in earlier stages. The greatest increase in the diversity of autotrophs is usually reached earlier in succession. The increase in diversity of species is believed to be a direct result of the increase in community biomass. As the organic makeup of a community enlarges, it develops zones or strata in which many new habitat niches are created. This stratification is especially evident in a forest.

Let us think of an oak-hickory forest as being divided into

layers, each of which provides habitats for different species of animals. As shown in Figure 21, the leaves and branches of the tallest trees make up the upper layer, called the canopy. Here we find many kinds of leaf-eating insects and insectivorous birds, like the red-eyed vireo. Squirrels nest there and feed on acorns and

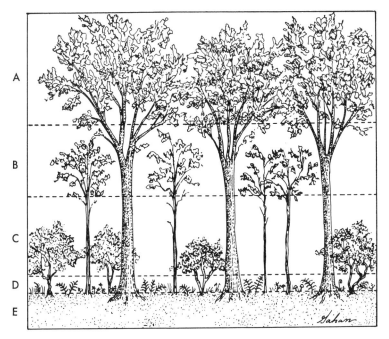

FIG. 21. *Stratification in a forest. (A) Canopy. (B) Understory. (C) Shrub layer. (D) Herb layer. (E) Litter and soil layer.*

hickory nuts. In the lower tree stratum, or understory, we find other insects, spiders, birds like the Acadian flycatcher, and mammals. Many of the animals move back and forth between layers. The Downy woodpecker, for example, moves up and down the trunks, extracting boring insects from the bark of trees. Below the understory is the shrub layer, which rises to about ten feet above the ground. It provides a habitat for still more insects, mice, and chipmunks. Certain birds nest in the thick shrub growth, from which they need travel only short distances to find food. The herb layer is made up of the smallest plants, rising a few inches above the ground. Here are many leaf-eating insects, spiders, toads, frogs, snakes, and ground birds such as the ovenbird. Rodents find cover under the herbs, and box turtles search for fungi, fruit, and small animals for food.

The litter and soil layer, making up the floor of the forest, contains a staggering amount of decaying material such as twigs, leaves, and waste products of animals. Over ten tons of leaves alone may collect on each ten acres of forest. Billions of fungi, bacteria, mites, springtails, pillbugs, millipedes, ants, and earthworms work in the litter and soil. As a result of their activities, more soil is formed and nutrients are cycled back to the plants in the forest.

The fourth trend deals with energy relationships. We learned in Chapter II that the total organic matter assimilated into the tissues of plants is called gross production and that part of its stored energy is lost, first in plant respiration and then in animal respiration. We also found that animals lose more of the stored energy in respiration than plants do. Relating this to succession, we note that in the early stages of succession, the loss of energy in the entire community will be relatively low because there are fewer animals to eat the plants. The gross production will therefore be greater than the community respiration ($P > R$), and the biomass and amount of organic material will accumulate. In the later stages of succession, as the number of animals increases and therefore community respiration also increases, losses from respiration will tend to catch up to gains in gross production and the net community production will decrease. Finally, in the climax, the annual production of organic matter equals total consumption ($P = R$), and all the energy available is used by the community to maintain ecosystem stability.

Climax Stability

Although the climax is relatively stable and long-lived as compared with earlier stages, it is not known if any community is completely self-perpetuating and permanent. Catastrophes such as storms, severe fires, and long periods of drought can shorten the life span of any community. For example, if a mature grassland is exposed to a series of dry years, it is set back to an earlier successional stage containing more annuals and short-lived perennials. There is some evidence that an aging process may be taking place in very old forests. Thus, young trees may not be replacing the old ones that die, and mineral cycling and energy flow rates may be slowing down. Some ecologists suggest that an old climax could eventually die and be replaced by a young community, perhaps different in species makeup. However, not much is known about this problem at the present time and more study is needed.

Significance of Succession

The climax community, with its larger biomass, greater diversity of species, and balanced flow of energy, is better able to cope with stresses and buffer the physical environment than a young community. Even though the climax is not as productive as earlier seral stages, its ability to survive in the face of a changing physical environment may be the primary purpose of ecological succession. Man seeks three things from the landscape: food, protection, and an aesthetic enjoyment of beauty. If we consider farmlands and pastures as modified grassland, then man must have earlier stages of succession as a source of food. But man needs the stability and beauty of the forest to give him a feeling of security and happiness. Such landscapes already exist in many places where agricultural fields are intermingled with forests on the hills and mountains. In cities, environmental groups recommend allowing most of the grass on the right of way of parkways to naturalize so that the expected new growth of trees might serve as a barrier against noise, headlights, and air pollution and also improve the beauty of the landscape.

in the winter. Precipitation is also low, but water is not a limiting factor because of the very low evaporation rate in the tundra. This biome, sometimes pictured as a "cold" desert, is better described as a wet, arctic grassland that is frozen most of the year. Although the ice in the top three or four inches of soil melts each summer to provide water for plant growth, a permanent layer of frozen ground, the permafrost, remains under most of the tundra and is responsible for the numerous lakes and bogs dotting the landscape.

The diversity of species in the tundra is relatively low, but a surprising number of plants and animals have evolved adaptations to survive the freezing temperatures. The thin plant cover is made up of very hardy lichens, mosses, grasses, and sedges. A few cold-resistant flowering herbs also survive as well as some shrubs, such as dwarf willows. These plants, supplemented by food from the Arctic Ocean and thousands of shallow ponds, support the food chains of the tundra. There is enough net primary production to feed thousands of migrating birds, emerging summer insects, and many mammals that remain active throughout the year. The mammals range from large animals, such as musk ox, caribou, polar bears, wolves, foxes, and marine mammals, to lemmings that tunnel about in the plant covering.

The cyclic irruptions and crashes of lemmings, snowshoe hares, and other populations in the Canadian tundra were discussed in Chapter V. The study of the tundra currently is getting a great deal of attention because its simplified structure makes it relatively easy for ecologists to investigate the basic makeup of an ecosystem.

Forests

Temperature and moisture are the two climatic conditions that determine whether or not a region will support the development of a forest. Trees are absent from the tundra, for example, because extremely low temperatures are limiting to trees. Just south of the tundra, however, warmer temperatures permit some forests to grow, and still warmer temperatures farther south allow other types of forests to grow. There is therefore a north-south temperature gradient along which we find three distinct forest types.

The northernmost forests, which form a broad band across Canada and Eurasia just south of the tundra, are characterized by cone-bearing trees, such as spruce, fir, and hemlock, that have evergreen needle-shaped leaves. Species diversity is low in these coniferous forests, and it is quite common to find pure stands of only one or two species.

In the more southern temperate regions, deciduous hardwood forests characterized by broad-leaved trees, such as oaks and hickories, that drop their leaves in the winter dominate the landscape. Although this biome includes a number of climax deciduous forests, such as beech-maple, oak-hickory, or oak-chestnut, all of them are climax temperate forests. The greater species diversity and stratification of these forests were described in Chapter VI. Deciduous forests once covered eastern North America and most of Europe, but logging and clearing for farms and cities have greatly reduced the extent of these forests today.

The tropical forests make up the third type, ranging from broad-leaved evergreen rain forests, where rainfall is abundant throughout the year, to tropical deciduous forests that briefly drop their leaves during a drier winter season. Two types of plants especially conspicuous in tropical forests are climbing vines called lianas and air plants called epiphytes that absorb rainwater through aerial roots as they hang suspended from the top branches of the canopy. Diversity of species is so incredibly huge in the rain forest that ecologists still do not completely understand this most complex of all ecosystems. It has been estimated that there may be more species of plants and insects in a few acres of tropical rain forest than in all the plant and animal life of Europe.

Two forest types may be said to follow a moisture gradient. One, the chaparral woodland described as a fire type in Chapter IV, develops in the winter rain–summer drought Mediterranean climate of coastal California. This forest is made up of dwarf trees, such as evergreen oaks, which are ecologically equivalent to similar plants in the "Mediterranean vegetation" of Spain, southern France, and Italy. The other, the moist-temperate rain forests along the coast from northern California to Washington, is made up of stands of Douglas firs and redwoods. These moist Pacific Northwest forests develop some of the largest volumes of timber in the world.

Deserts

Deserts occur in regions with less than ten inches of annual rainfall, or in hot regions where there is more rainfall but it is scattered unevenly during the year and the rate of evaporation is high. Two kinds of deserts can be identified in North America: the "cool" type in Washington, with sagebrush and cheat grass, and the "hot" type in Arizona, with creosote bushes and cacti. The "cool" deserts often lie in the dry rain shadows on the leeward slopes of high mountains, which block off moisture from the sea. Conditions of subtropical temperatures, steady wind, high evaporation, and rain in winter are responsible for producing the "hot" deserts of the Southwest. Closer to the equator, these same conditions produce the great deserts of the world, ranging from the Sahara in North Africa through Arabia and Iran to the great Mongolian Desert.

Four kinds of plants show specific adaptations for life in a desert. First, there are the shrubs, such as creosote and sagebrush, which shed their short, thick leaves during dry spells and become dormant before wilting occurs. The characteristic wide spacing between plants was explained in Chapter V. Second, there are fleshy plants, such as the cacti of American deserts and the euphorbias of African deserts, which store water in their tissues. Third, there are annuals, such as cheat grass and wildflowers, which grow swiftly after a rain, bloom, produce seed, and die, thus avoiding drought. Fourth, there are very small plants, including mosses, lichens, and blue-green algae, which lie dormant in the soil and become active when moisture is present.

Animals of the desert such as insects and reptiles can survive on metabolic water produced in the respiration of carbohydrates because their chitinous or scaly skin and dry excretions effectively conserve body water. Kangaroo rats and other nocturnal rodents that excrete a concentrated urine and do not sweat can live entirely on seeds without drinking water. The camel can go for five to twelve days without drinking water because it can store water in its tissues. Predators like the roadrunner bird are active during the early morning and late afternoon, and others like the kit fox and coyote search for prey during the night.

Grasslands

Grasslands usually occur where the annual rainfall is intermediate between that of the desert and that of forests. In the temperate United States, for example, twenty to thirty inches of rain would support a short grass prairie good for grazing, and forty inches would allow development of a tall grass prairie where wheat and corn might be grown. Grasslands usually occupy large areas of the interior of continents and are known by such local names as the Great Plains in the United States, the steppes in Russia, the veldt in South Africa, and the pampas in South America.

The dominant plants are grasses, which range in size from tall species reaching five to eight feet to short ones reaching a height of six inches or less, growing in clumps or spreading uniformly by means of underground rhizomes. Nongrassy flowering herbs, called forbs, are often important components of plains and prairies and are eagerly eaten by all livestock. Trees and shrubs may also be present, either scattered as in tropical savannahs or in groups along streams.

The soil built up by a grassland community is very different from the soil made in a forest. When dead grass remains in a grassland, the first phase of decay is rapid, producing much humus and little litter, but the second phase, the recycling of humus into minerals, proceeds at a slower rate. As a result, a dark soil develops that has five to ten times as much humus as forest soils.

Large grazers of the grasslands include running types like bison and antelope and burrowing types like ground squirrels and gophers. When man uses grasslands as pastures, he replaces the native grazers with his domestic cattle, sheep, and goats. Predators in American grasslands, including coyotes, bobcats, kit foxes, badgers, hawks, and owls, feed principally on burrowing rodents, but bands of coyotes may sometimes prey on weaker or younger members of a grazing herd. The large predators in the African savannah, such as lions and cheetahs, tend to hunt singly or in family groups; jackals and hyenas, on the other hand, always prey in packs. Because of their huge size, elephants and rhinoceroses are seldom attacked.

Aquatic Biomes

The climatic factors that helped us classify the terrestrial biomes are not applicable to large bodies of water, where plants do not exert the controlling influence that they do on land. Since lakes naturally succeed to land biomes, we will consider lakes and streams first and then study the largest of the biomes, the sea.

Ponds and Lakes

Ponds and lakes are aquatic ecosystems of standing fresh water, most of the larger ones dating back to the ice ages. They are usually fed by groundwater, which balances the water lost by evaporation. As shown in Figure 24, three zones can typically be distinguished in lakes and large ponds: a littoral zone containing

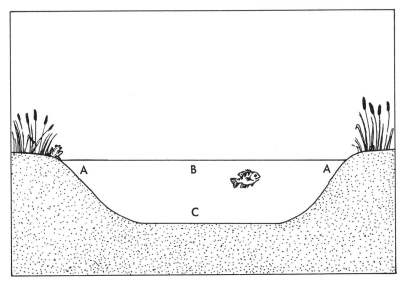

FIG. 24. *Major zones of a typical lake. (A) Littoral zone—along the shore. (B) Limnetic zone of open water. (C) Profundal or deep-water zone.*

rooted plants along the shore, a limnetic zone of open water dominated by phytoplankton, and a deep water or profundal zone containing only heterotrophs. The rooted plants of the littoral zone, which include cattails, waterlilies, and pondweeds, provide a habitat for many kinds of aquatic insects, worms, and snails. Fish and other large animals, such as frogs, feed on the smaller ones. The pondweed potomogeton is relished and eaten by all kinds of waterfowl and mammals such as the muskrat. The primary producers of the lake ecosystem are the floating algae making up the phytoplankton of the limnetic zone. They include diatoms, desmids, blue-green algae, and green algae. They are eaten by a zooplankton of protozoa, rotifers, and crustaceans such as daphnia. Fish such as minnows, bass, and perch are able to swim freely through all levels of the lake. Some spend the winter in deeper waters and move to the surface in the summer. The hetero-

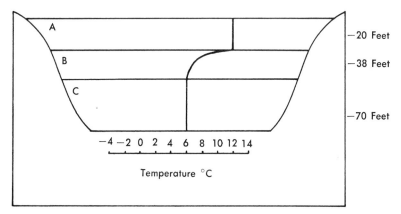

FIG. 25. *Thermal stratification of a lake in summer. (A) Warm epilimnion (12° C.). (B) Thermocline zone—gradual drop in temperature of 1° C. for every 3 feet of increase in depth. (C) Hypolimnion has uniform temperature of 6°C.*

trophs in the deep-water zone feed mostly on detritus and dead matter falling down from above. Among them are the decomposers and dwellers such as worms, clams, insects, and crayfish.

Lakes in temperate regions may also become thermally stratified in summer and again in winter because of unequal heating and cooling. Figure 25 illustrates summer stratification in a 70-foot-deep lake in which the upper layer, the epilimnion, is warm (12° C) compared to the lower layer, the hypolimnion, which has a uniform temperature of 6° C. The two layers are separated by a thermocline zone in which there is a gradual drop in temperature of 1° C for every 3 feet increase in depth. During winter stratification, the epilimnion is a layer of ice at 0° C, floating on top of the thermocline, which separates it from the warmer hypolimnion, which has a uniform temperature of 4° C. This happens because water reaches its maximum density at 4° C and when cooled further expands and becomes less dense. It expands even more at 0° C, when it freezes into ice and floats on the top of the lake.

The thermocline acts as a barrier to the exchange of materials. As a result, oxygen in the hypolimnion and nutrients in the epilimnion may run short during summer and winter. In the spring and fall, however, as the entire body of water approaches the same temperature, mixing occurs among the layers and oxygen and nutrients are circulated to all parts of the lake. Blooms of phytoplankton often follow these seasonal overturns that rejuvenate the ecosystem.

Depending on their productivity, lakes are usually classified as oligotrophic (low productivity) and eutrophic (high productivity). To determine the productivity of a lake, a number of measurements are made. One is the weight of the total suspended solids found in a given volume of water. These suspended materials include the phytoplankton, zooplankton, silt, human sewage, and wastes from animals, plants, and industry. The ecologist can then calculate how much life, such as the number of pounds of fish per unit volume, the lake can support.

When agricultural land is scarce, as in Japan, aquaculture has been practiced as a supplement to farming for many years. In this method of fish culture, seminatural bodies of both fresh and salt water are skillfully managed to produce large yields of algae, fish, and shellfish. Compared to an estimated production of fish in natural oceans of from 5 to 100 pounds per acre per year, some of these ponds have yielded as much as 1,000 to 5,000 pounds per acre per year of such herbivorous fish as carp, for example. This compares very favorably with the 250 pounds of cattle produced on grazing lands. Even greater yields have been reported as a result of improved techniques in silo fish farming, a growing industry in the United States and some other countries.

Freshwater Marshes

As described in Chapter VI, a freshwater marsh represents a seral stage in lake succession leading eventually to a climax forest in temperate regions. Like estuaries, they tend to be naturally fertile ecosystems. Although tides are absent, periodic changes in water levels from increased rainfall help to keep the marsh stable and fertile. During dry periods, fires help to maintain the marsh and retard ecological succession by deepening its water-holding basin and decomposing accumulated organic matter.

Freshwater marshes support an incredible variety of wildlife. For example, the wetlands of the Florida Everglades are habitats for such animals as marsh rabbits, deer, panthers, bears, herons, egrets, alligators, rattlesnakes, turtles, amphibia, and countless insects. Marshes also provide seasonal homes for millions of migratory birds. It is estimated that 80 percent of American waterfowl breed in the wetlands of the northern central states, Canada, and Alaska. In addition to producing wildlife, marshes are important in maintaining the water tables in nearby ecosystems. Rice paddies represent artificial marsh ecosystems that produce a high yield of rice, a cultivated marsh grass.

Streams and Rivers

Rapid movement of water makes conditions in streams quite different from those in lakes. The turbulent running water not only makes oxygen abundantly available but also keeps the stream's temperature low. In addition, the constant turnover of the water keeps the temperature uniform from top to bottom.

Conditions change along the length of a stream as we go from the cold, swift flowing, clean liquid of its headwaters to the warmer, slowly moving, muddy waters of its mouth. In many cases, these situations alternate in the same stream and we find rapids or riffles with hard, rocky bottoms and pools with muddy bottoms. Because of the different conditions of existence, communities in rapids are very different from those in pools.

The organisms in rapids are uniquely specialized to withstand the swift flow of the water. Its producers are algae and mosses that can attach themselves to the bottom rocks. Many kinds of aquatic insects, such as the nymphs of mayflies and stoneflies, are able to exist here because they have flattened, streamlined shapes, strong appendages, and suction discs that enable them to adhere to the undersurface of rocks. Some caddis fly larvae build weblike nets that are attached at one end and suspended in the stream to remove food particles from the flowing water. The fish in fast streams, such as trout and minnows, have powerful muscles that enable them to swim against strong currents.

In the slower moving waters of pools, the communities tend to be very similar to those in lakes. There may be a good development of plankton organisms, aquatic insects, and fish. The rich supply of organic debris settling to the bottom supports a large community of decomposers and scavengers. Tubifex worms thrive in oxygen deficient conditions that may develop in the muddy bottoms. An increase in the tubifex population is often an indication that the oxygen level is falling.

The Sea

The sea covers 70 percent of the earth's surface and contains the greatest abundance and diversity of organisms in the world. It is the thickest layer of living things in the biosphere, averaging between 2 to 3 miles in depth. The makeup of its biological communities is largely determined by dominating physical factors, such as salinities, temperatures, light intensities, waves, tides, currents, and pressures. Its salt concentration may be expressed as 3.5 percent, of which 2.7 percent is sodium chloride. The remaining 0.8 percent consists of other salts, such as calcium carbonate and magnesium sulfate. Because salinity lowers the freezing point of water, it does not freeze at 0° C and continues to become more dense as it cools until the freezing point is reached. Bottom water is near 0° C in most places. The temperature of sea water is fairly constant, almost all of it being colder than 7° C except for a thin surface layer and the shallow water of bays, which do not go higher than 30° C. For convenience of description, the ocean biome is divided into two smaller zones: the continental shelf water that extends to a depth of 600 feet, and the deep sea water beyond. Since the maximum depth to which light penetrates in the sea is 600 feet, the water of the deep sea zone below this depth is in perpetual darkness.

Continental shelf water is fairly productive, especially where upwelling provides the phytoplankton with a good supply of phosphates. Filter feeders, like sardines and anchovies, strain plankton from the water through their gills, and salmon, mackerel, and tuna feed on them and other organisms. The bottom mud dwellers, still in contact with the plankton, include many burrowing animals, such as clams, snails, worms, and shrimp. They are eaten by crabs, lobsters, starfish, and a number of different kinds of fish, including flounder, rays, haddock, halibut, and cod. In places where the bottom is rocky, seaweeds, sea anemones, and corals may be found attached to the hard surface. The sea food harvested in this continental shelf zone is an important source of proteins and minerals for humans.

The vast stretches of the deep sea zone are semideserts by comparison. The distribution of phytoplankton is patchy and limited by a short supply of nutrients. Even though man is not able to get much food from the deep sea zone, it is still very important to him because its large area helps to regulate land climates and maintain favorable levels of oxygen and carbon dioxide in the atmosphere. The deep sea zone also serves as a storehouse for valuable minerals lost from the continents.

In the dark, cold region of the deep sea, producers are absent and small animals feed on organic debris falling down from above. Many deep sea fish like the angler have curious adaptations, such as luminescent organs that produce their own light. They also have huge mouths and stomachs, which enable them to swallow bulky animals that are scarce and hard to find.

Coastal Ecosystems

There are a number of ecosystems between the sea and the coast that are part of what might be called an intertidal zone. Although conditions of temperature and salinity are more variable here, food conditions are so superior that the area is packed with living things. The organisms that dwell in these ecosystems are exposed to sunlight, air, and land predators such as shore birds at low tide and to salt water and ocean predators at high tide. In addition, they must be able to survive the pounding of the surf during much of the day. Four examples of coastal ecosystems are rocky shores, sand beaches, intertidal mud flats, and estuaries.

A rocky shore is thickly covered with seaweeds such as fucus and laminaria and animals such as periwinkle snails, mussels, oysters, limpets, and barnacles that are firmly attached to the rocks. They are protected from extremes of temperature and dryness by tightly closing shells. Tidepools, found in depressions in the rocks, are inhabited by crabs, snails, and sea anemones.

A sandy beach is likely to have worms, clams, snails, and crabs that can burrow beneath the shifting sand. Similar denizens also occupy burrows in intertidal mud flats when the tide is out. These animals are of two general feeding types: filter feeders, such as clams, that filter out food particles from water, and deposit feeders, such as snails, that extract organic matter from the mud they ingest into their digestive tubes. Part of the food is produced by algae living on and in the sand and part is carried in with the tides.

An estuary (from the Latin *aestus,* meaning "tide") is a tidal marsh area, or wetland, found in coastal bays and mouths of rivers where the salinity is intermediate between the sea and fresh water and where tidal water is an important regulator. An airplane view of an estuary discloses networks of tidal creeks running through vast areas of salt marshes.

The shallow creeks and salt marshes are among the most naturally fertile areas in the world. They not only support a luxurious community of stationary organisms but also serve as breeding grounds for fish and shrimp that later move back to the open sea. One reason for the high energy flow rate is the tidal action that provides a rapid circulation of nutrients and food and helps remove metabolic waste products. Another reason is the diversity of autotrophs that work together to maintain a high primary productivity. These autotrophs include phytoplankton, algae living on and in the bottom sediments, submerged attached seaweeds, eel grasses, and the salt marsh grasses.

VIII

Man and the Environment

We began our study of ecology with a brief description of the environmental crisis and pointed out a number of extremely serious problems that demanded immediate attention. Workable solutions to these problems, based on a deep understanding of ecological principles and man's role in the balance of nature, have been proposed by leaders in the environmental movement and are being pushed politically, legislatively, and legally. Our final chapter is devoted to an overview of these solutions, which often involve high costs, political adjustments, and changes in lifestyles.

Air Pollution

We use up 18 quadrillion kilocalories of energy per year in heating our homes, producing electricity, powering our factories, running our automobiles, and flying our airplanes at high altitudes. Most of this energy comes from the combustion of fossil fuels—oil, natural gas, and coal. Formed millions of years ago from the organic remains of ancient organisms, they are found today in limited quantities as deposits in sedimentary rocks. Estimates show that 43 percent of our energy comes from oil, 33 percent from gas, and 19 percent from coal. Only 5 percent comes from other sources—hydroelectric, 4 percent, and atomic energy, 1 percent. Under ideal conditions of complete burning, the carbon in coal and the hydrocarbons in oil and gas combine with atmospheric oxygen to produce carbon dioxide and water vapor. If insufficient air is supplied or the temperature is too low, unburned hydrocarbons and poisonous carbon monoxide will be present in the gaseous products. If the temperature is very high, oxides of nitrogen will result from the reaction of atmospheric nitrogen with oxygen.

Hydrocarbons and nitrogen oxides are particularly important in the production of photochemical smog, typical of Los Angeles. In the presence of sunlight, a very complex reaction produces ozone (O_3) and a by-product, peroxyacylnitrate, which can damage crops and cause tearing of the eyes and irritation of the mucous membranes in humans.

In places like London and New York City, a temperature inversion may cause a layer of cool, foggy air to be trapped near the surface of the earth under a layer of warmer air. When this occurs, the pollutants poured into the air from chimneys and automobiles cannot escape into the upper atmosphere and a smog develops (smoke + fog = smog). Under such conditions, air pollution is extremely hazardous and emergency measures to stop all pollution may be called for.

Sulfur dioxide, an invisible gas that in high concentrations can make breathing difficult and even cause death, is the most serious pollutant in large cities. All of it comes from fuels containing sulfur, such as coal and high-sulfur oil. In New York City, through tight enforcement of the law requiring 99 percent sulfur-free fuels, 56 percent of the 1966 amounts of sulfur dioxide have been eliminated. Particulate matter, unburned particles of carbon (soot) and ash, can be reduced by requiring that all oil burners and incinerators be upgraded with modern controls, and that operators of such equipment be trained in air-pollution control techniques.

The automobile is the most important source of pollution in our cities, and increasing attention is directed to its emissions—carbon monoxide, hydrocarbons, oxides of nitrogen, and lead. As mandated by the 1970 Federal Clean Air Act, the Federal Environmental Protection Agency has set the end of 1976 as a deadline for the cleaning up of the air in New York City and nine neighboring New Jersey counties. In New York City, tolls may be imposed on all East River and Harlem River bridges to ease midtown traffic and encourage the use of mass transit. There will be an annual emission inspection of all passenger cars and installation of fume control devices on heavy duty vehicles and twice yearly inspection of them. All 1975 and later model cabs will also need fume controls and be given thrice yearly exhaust inspections.

Manufacturers of aircraft engines will be required to meet federal air-pollution emission standards by 1979. These standards will result, for jet engines of more than 8,000 lbs. thrust, for example, in reductions of 60 percent for carbon monoxide, 70 percent for hydrocarbons, and 50 percent for nitrogen oxides. Since industrial air pollution approximates 10 percent of the total air pollution in a city, all industrial control apparatus must be inspected to ensure compliance with emission standards.

To provide the information on which emergency measures may be taken during times of severe pollution, it is necessary for a city to maintain a monitoring system in which pollution levels are measured and recorded daily. An example of satisfactory levels would be:

Sulfur Dioxide	.06 ppm (parts per million)
Carbon Monoxide	2.00 ppm
Smoke Shade	.06 units

We are faced today with an energy crisis precipitated by diminishing supplies of oil. This crisis calls for increased funding by electric utilities for research and development into means of conserving fuel and producing electricity without adverse effects on the environment. Environmentalists recommend research into the use of solar energy as well as the use of total energy systems that generate power from waste heat created by the generator itself. High savings can also result from changes in the design of appliances that receive constant use, such as refrigerators and air conditioners. There is also a need for researching new energy sources for automobiles. But in the meantime, concern over the energy crisis must not be used as an excuse to abandon such environmental controls as sulfur limits on fuels and emission control devices on cars.

Municipal Sewage Pollution

Because air pollutants are not normally found in nature, there are no organisms capable of using them and breaking them down into harmless forms if they reach excessive concentrations. The organic matter in sewage, on the other hand, is normally present in water ecosystems. Thus, there are adapted microflora available that can break the organic stuff down into simple compounds needed by phytoplankton. Too much sewage, however, can cause such an overpopulation of decomposers that most of the dissolved oxygen is used up in their metabolism and very little is left for the fish and other organisms. When the dissolved oxygen becomes too low to support life in a stream or lake, only anaerobic bacteria and pollution-tolerant animals, such as sludgeworms (tubifex), may be left. It is therefore necessary to remove most of the organic material in sewage treatment plants before discharging the sewage into a body of water. Although in the last fifteen years the federal government, states, and cities have spent $5 billion on federally aided construction of local sewage treatment plants, the sewage of half of the nation's population is still inadequately treated.

One of the current national goals is to provide secondary treatment to at least 90 percent of the municipalities within the next five years at a cost of over $10 billion. In primary treatment, after debris is removed by screens, the sewage then passes through sedimentation tanks where the solid materials settle to the bottom and become sludge. This treatment removes only about one third of the impurities. Secondary treatment passes the water over a bed of rocks 3 to 10 feet deep where decomposers reduce the organic load by 90 percent or more as the water trickles through the rock filter. The treated sewage, after chlorine is added to kill most of the disease germs, is then discharged into a stream, lake, or ocean, where the small remaining organic matter can be safely absorbed. Advanced tertiary systems also remove phosphates and nitrogen and improve the odor and taste of the water.

Sludge is disposed of in a number of ways. Communities that border on an ocean dump the sludge at specified sites, usually about 12 miles offshore. The Federal Environmental Protection Agency is presently planning to phase out such sites and stop all ocean dumping by 1981 according to an orderly and environmentally accepted plan. Some inland cities may send half of their sludge to agricultural areas for use as fertilizer and bury the other half in the ground.

Industrial Water Pollution

The contamination of the U.S. water supply by industry is our major water pollution problem. As much as 40 percent of the water processed by municipal sewage plants could be effluent from factories. In addition, a large amount of untreated waste water is dumped directly into streams, polluting our water supply and the oceans with an amazing array of chemicals and radioactive substances. A group of chemicals called PCBs (polychlorinated biphenyls), used in industrial processes to improve rubber, plastics, and paper products, has been found to cause embryonic deaths in the eggs of birds. Lead, which is added to gasoline as an anti-knock ingredient, is increasing in concentration in surface waters of the continental shelf and reaching levels that may be toxic to phytoplankton, the producers of 75 percent of the world's oxygen. Mercury, used in the paint and plastics industries, recently made headlines when the Food and Drug Administration advised the public to stop eating swordfish because 95 percent of the samples examined were contaminated with this poisonous metal. Oysters have largely disappeared from bays where industrial wastes and sewage have been dumped.

In October 1972, Congress passed the Clean Water Act, which aims to end the pollution of United States waterways by 1985. Every pollutant is defined and limited. By July 1, 1977, industries would have to install the "best practicable" anti-pollution devices on all their waste systems. By 1983, all industries must upgrade their anti-pollution equipment to make use of the "best available" technology even if it is expensive.

Oil has become one of the most common ocean pollutants. Oil tankers at sea have leaks and spills that kill large numbers of sea birds, fish, and other marine life and blacken miles of beaches. In addition, oil leaks from offshore drilling equipment and from land refineries near rivers add to the pollution of coastal waters and beaches. International laws to regulate oil discharges have been established by many nations.

Detergents in city and industrial sewage have added to the pollution problem in recent times. Some of the earlier cleaning chemicals could not be broken down by decomposers and formed blankets of white foam over the water surfaces of streams and lakes. Manufacturers replaced these cleaning chemicals during the 1960's with biodegradable materials that could be broken down by bacteria, and the blankets of suds disappeared. But a high percentage of phosphates still remained to bring on huge blooms of algae that grew so thick in the surface waters that they cut off sunlight and oxygen. As these masses of algae died and sank to the bottom, decomposing bacteria increased in large numbers and finally reduced the oxygen supply below the level needed to support aquatic life. This process of eutrophication speeds up the natural succession that eventually changes a lake into a marsh. Legislation barring the sale of phosphates is now in effect in a number of states. In New York State, for example, the law reduced the allowable amount of phosphates in detergents to 8.7 percent in 1971 and prohibited the sale of any household detergents containing more than a trace of phosphates after June 1, 1973.

The use of water from a river, lake, or ocean to cool the condenser of a nuclear or fossil-fueled power plant poses a threat to the fish and other organisms living in these bodies of water. The striped bass of the Hudson River are endangered by the three nuclear power plants built by Consolidated Edison at Indian Point. Cooling water is obtained by pumping water from the river into an intake pipe, passing it through the condenser and returning it to the source through an outlet pipe. This kills the fish in three different ways: (1) entrainment, or sucking in of tiny planktonic organisms, including embryonic stages of fish; (2) impingement, or squashing of young fish on the intake screens; and (3) thermal pollution, in which the heated water, raised 15° F in temperature, disrupts life cycles and spawning behavior. Conservation groups are urging that cooling towers be installed to cool the heated water through an open air evaporation and cooling system so that the same water can be used over again. With a reduced need for water, the pumps can operate with much less force and pull fewer organisms through the intake screens.

Agricultural Water Pollution

The use of chemical fertilizers in agriculture has contributed to the pollution of our waters. Rich in nitrates and phosphates needed by plants for synthesizing tissue building proteins and nucleic acids, fertilizers are added to soils to make up for the loss of these nutrients in the harvesting of crops. Through their use, farmers have achieved higher crop yields than ever before. But chemical fertilizers create pollution problems when the water run-off from the fields transports them into streams and lakes, where they result in an abnormal growth of algae. The consequences are the same as those described for phosphate detergents. In the ocean, fertilizers may sometimes cause the fish-killing red tides produced by a population explosion of dinoflagellates.

A possible solution to this problem is the substitution of composting for fertilizing, as practiced in organic farming. In this process, systematically layered organic waste materials are reduced to a rich humus that both conditions the soil and acts as a natural fertilizer. Rotation of crops, described in Chapter III, restores nitrates to the soil.

A more serious threat to the ecological balance of the oceans is the use of pesticides to control crop destroying insects. These poisons are sprayed over cultivated fields, often in broadcast fashion, at frequent intervals during the growing season. Like the fertilizers, these insecticides find their way into bodies of water as a result of normal runoff from the land. However, unlike the fertilizers, many of these chemicals are persistent and do not break down into simpler, harmless substances. As a result, they are passed along in the food web, becoming more and more concentrated at each succeeding trophic level.

The best known of the pesticides, DDT, a chlorinated hydrocarbon developed during World War II to control typhus fever transmitted by body lice, was widely used after the war to check crop eating insects. After about thirty years of use, we now find that DDT is threatening the survival of fish eating birds like the osprey, brown pelican, eagle, and peregrine falcon. Concentrated DDT in the mother bird interferes with the production of calcium; this results in thin-shelled eggs with a high incidence of breakage. Ecologists have determined that the "safe" level of .00003 ppm of DDT that enters Long Island Sound increases to .04 ppm in the zooplankton. The DDT level rises to .5 ppm in the minnows that feed on the microscopic animal life and, in turn, reaches a concentration of 2 ppm in the larger fish that devour the minnows. Finally, ospreys and cormorants eat the fish and the pesticide level reaches 25 ppm, an increase by a factor of 10 million over the "safe" concentration that enters Long Island Sound.

Biological control is recommended as a solution to the pesticide problem. In one method, interplanting of crops may be used to control insects. In another, insects introduced from other parts of the world are brought under control by importing the natural predators or parasites that regulated them in their original habitats. In the first application of this type of control, ladybird beetles were introduced in 1888 from Australia to control the cottony cushion scale in the citrus groves of California. Japanese beetles were controlled by introducing a fungus disease that attacks the larvae. Organic farmers can order praying mantises and ladybird beetles by mail from organic suppliers.

Another type of biological control is the sterile male technique introduced in the 1950's to control the screw-worm fly whose larvae live in the flesh wounds of sheep and cattle. The male flies are rendered sterile by exposure to radiation from cobalt-60, which, however, does not affect their sexual aggression. They are then released from airplanes in numbers larger than those of the males in the natural population, so that females copulating with them proceed to lay unfertilized eggs. Screw-worm flies were completely eliminated from Florida in a few months by this technique.

The IBP program seeks to find biological controls for four pest groups—aphids, spider mites, scale insects, and rice pests—all of which destroy agricultural crops around the world. This research is closely coordinated with those in other IBP countries working on insect pests.

Solid Waste Disposal

The growing volume of solid wastes and the paucity of recycling programs have created a number of problems for our cities. For example, New York City produces 30,000 tons of waste a day and is running out of places to put it. To solve this problem, New York's remote disposal program proposes to pay contractors $8 million to $10 million a year to bale and ship 730,000 tons of solid waste annually to sites, usually scarred by strip mining, outside the state. Although this plan helps to solve the problem of unsightly abandoned strip mines, many citizens object to their town's being known as the garbage dump of New York City, or any other city.

In the United States as a whole, 77 percent of all solid wastes are dumped in open country, 13 percent is buried in landfills, and 10 percent are burned in incinerators. These disposal procedures are unsatisfactory because dumps and landfills breed armies of rats, and insects and incinerators contribute to air pollution. In addition, experts estimate that a billion dollars' worth of reusable materials is wasted every year by such methods of disposal.

A number of recycling programs have been instituted on a small scale in many communities. For example, newspapers, bundled and tied, have been redeemed at 35 cents per hundred pounds and turned over to the Recycling Corporation of America. To further this effort, the federal government is now buying at least half of its paper from recycled stock. Environmental groups are urging telephone companies to recycle their telephone books and newspapers to reuse some of their old editions.

The million bottles and million and a half cans that are tossed away every day in our big cities can also be reclaimed. Several glass manufacturers now pay for throwaway bottles, and the Reynolds Company buys used cans and other scrap aluminum. Taking the lead from Oregon and Vermont, the New York legislature is now considering a bill that prohibits the sale of non-reusable beverage containers. It is also considering a bill to stop the abandoning of old cars by requiring a disposal bond of $50 to be paid to the State Commissioner of Motor Vehicles upon the purchase of an automobile.

Development of resource recovery plants will go a long way toward solving our waste disposal problems. A mechanized refuse processing plant that can process 150 tons of trash a day is already in operation in the city of Franklin, Ohio. From that it will recover some 27 tons of paper fiber, 9 tons of ferrous metals, 9 tons of glass cullet, and 1,500 pounds of aluminum as well as recycle a large quantity of compostible residue for agricultural use.

Noise and Sight Pollution

The background din to which people are subjected is estimated to have doubled in the last generation. The World Health Organization says that noise costs the United States $4 billion a year in accidents, absenteeism, inefficiency, and compensation payments. Only a few states have gone as far as setting theoretical limits on automobile noise, and these are scantily enforced. Regulations limiting slower jet planes to tolerable noise levels are in effect or pending. One reason why Congress rejected a program to develop SST's (supersonic transports) in 1971 was because they produce noisy sonic booms.

New York State's Environmental Protection Bureau investigates complaints relating to excessive noise from discotheques, outdoor speakers, and condensers of air conditioning systems. The complaints are investigated and steps are taken to reduce or eliminate the noise.

Another pollution problem is the spread of billboards, junkyards, and the general unsightliness of urban areas. State laws require junkyards to be fenced so that they are hidden from public view. Educational campaigns are needed to combat littering.

Stripping the Landscape

A million acres of green plants are paved under each year, thus reducing photosynthesis and the renewal of oxygen in the air. These areas are covered with highways, roads, parking lots, railroads, airports, and asphalt and concrete surfaces. Strip mining for coal has gouged out a number of landscapes, removed their green covering, and left them exposed to erosion. What is needed is an enlightened control of productive economic activities.

Estuaries or tidal marshes, recognized by environmentalists and legislators as the last and most important of the natural resources of the East Coast, are now protected by wetland acts in New York, Connecticut, and New Jersey. The 1973 Tidal Wetlands Preservation Act in New York, for example, requires that a state permit be issued before an alteration can be done on wetlands, even if the land is privately owned. A number of private owners, who have been denied permission to develop homes on their salt marshes, feel that they should be compensated for the lands as stated in the Fifth Amendment to the U.S. Constitution: ". . . . nor shall private property be taken for public use without just compensation." Several scientists, basing their calculations on the flow of energy produced by an acre of wetlands, have recently translated the worth of an acre to be roughly $4,100 a year.

Because of the ruthless cutting down of forests to produce farmlands, except in a few protected places, most of the great forests that our pioneers found are gone. The goal of ecological forestry is to maintain the integrity of forest lands, protect their wildlife, secure their recreational and aesthetic values, and at the same time permit harvesting of their great renewable timber resources. Clear cutting, in which an area is completely cleared of trees, is, in most cases, the direct opposite of ecological forestry. Selective cutting of mature trees on a sustained yield basis is a more acceptable method of lumbering.

Overgrazing and careless farming have resulted in the loss of the plant cover and erosion of the topsoil. Right into the present century the practice of many farmers in the United States was to move on when the soil could no longer produce any crops. Farmers planted the same crop year after year, and few efforts were made to use organic fertilizer to restore the fertility of the soil. Poor farming methods caused the loss of rich topsoil and led to the creation of the Dust Bowl and other deserts. We can conserve our precious topsoil through improved farming methods that include contour plowing, terracing, strip planting, and rotation of crops.

The use of irrigation in arid areas, although producing short-range agricultural benefits, often leads to long-range dangers such as salinization. As the large volume of water passes through the irrigation system, salts may be left behind, accumulate over the years, and eventually limit the growth of crops. Many ancient irrigation systems of the Middle East, Greece, China, and India are filled with silt, the end product of soil erosion. In the irrigated lands of the western United States, there is a constant danger of salinization and means of avoiding this difficulty need to be devised.

Upsetting the Balance of Nature

Man often upsets the balance of nature when he attempts to protect animals by controlling their predators. For example, as pointed out in Chapter V, when the natural predators of deer are removed, deer overgraze their forest habitat and many die from starvation during the winter. When hawks are killed to protect chickens, rats increase in numbers and cause extensive damage to crops.

In good wildlife management, game birds such as quail are protected from hawks by providing vegetative escape cover near the feeding areas. This cover enables the quail to easily elude the talons of a hawk.

Problems are often created when animals are introduced into new regions without the regulation of their natural predators. In one famous example, six pairs of rabbits, imported into Australia in 1859 and released on an Englishman's estate to breed a hunting stock, multiplied so rapidly that they now number in the hundreds of millions and do untold damage to crops and pasture lands. Similarly, 100 starlings, imported into Central Park from Germany in 1890, started spreading in the 1920's and now occupy the greater part of the eastern United States, where their gregariousness makes them extremely unpopular.

If a predator is introduced to control a pest, it should be specific for the species to be controlled or it may itself become a pest. For example, the English sparrow, introduced into New York City to control the elm spanworm, not only failed to control the insect but became a pest throughout most of the United States and Canada. In like manner, the mongoose, a small weasel-like predator introduced into the islands of Puerto Rico and Jamaica to control rats, found it more convenient to prey on ground nesting birds.

The Welland Canal, connecting the St. Lawrence Seaway to the Great Lakes, let sea lampreys and alewives enter the upper Great Lakes with disastrous effects. By the mid-1950's, trout, the backbone of the Great Lakes fisheries, were almost completely wiped out by the parasitic lampreys, and the alewives increased in such large numbers that many of them died and were swept up on bathing beaches.

Congress approved the construction of an 800-mile pipeline across Alaska in 1973 despite the objections of such groups as the Wilderness Society, Friends of the Earth, and the Environmental Defense Fund, which feared that the 4-foot-wide pipe would damage the Arctic tundra and lead to oil spills in coastal waters. The legislature felt that it was more important to reduce current oil shortages and lessen America's dependencies on imported petroleum from the politically unstable Middle East. In another case now under consideration by the Council on Environmental Quality, environmentalists, conservationists, and a whole range of legislative and governmental figures are lined up against the oil industry in opposing the leasing of the Atlantic Coast for offshore drilling because they are skeptical that offshore development can take place without irreparable harm to the ecology.

Conservation of Wildlife

Lists of endangered species have been compiled by the International Union for the Conservation of Nature, which has its headquarters at Morges, Switzerland. Its "red data book" lists mammals, such as the giant panda, and large birds, like the California condor, that are in danger of extinction. The "black data book" names species that have been lost since 1600. Of the 4,226 species of mammals that were alive in 1600, 36 have become extinct and 120 are in danger of extinction. For birds, 94 are extinct out of 8,684 species in 1600 and 187 are now threatened.

A species can become extinct when man kills it directly by unrestricted shooting, poisoning, or trapping. Most North American wildlife is well protected from indiscriminate shooting by state and federal laws, refuges, and a growing public awareness of value and beauty. Poaching and illegal hunting of protected species are still major problems, but the violators risk heavy fines or jail. Many species are doomed by land clearing and settlement in their major nesting grounds. Most of the older nations of the world have well-financed programs to protect remnant wildlife populations in natural parks, refuges, and game reserves. The smaller and newer nations will need the assistance of international agencies such as the World Wildlife Fund, which provides money to existing scientific or cultural organizations to meet emergencies in wildlife conservation.

In addition, cities and states have local regulations concerning the care and protection of animals. New York State's Mason Law protects endangered species by prohibiting their sale, or the sale of products made from their skins or furs. A few of the animals included are the crocodile, alligator, wolf, leopard, jaguar, tiger, polar bear, and caiman. New legislation is being considered to improve conditions and treatment of animals in pet stores, private zoos, motion pictures, and transportation.

A problem currently facing developing nations in Africa is whether or not to exterminate the great variety of grazing mammals found in their grasslands and substitute cattle as a means of providing more complete proteins in the diet of African people. Many ecologists suggest sustained harvesting of the antelopes, hippopotamuses, wildebeests, and other grazers because not only are these diverse species able to make the widest use of the grassland's food production but they are also naturally resistant to tropical diseases to which cattle are vulnerable. A number of ranchers in Africa, operating under special permits from their governments, have demonstrated that raising wildlife under natural conditions is more profitable than conventional farming.

The Population Explosion

One of the major problems facing the world today is providing food for a human population that increases at a compound interest rate of 1.7 percent a year. At this rate, the world's population of 3.6 billion persons will double in thirty years. The United States alone will grow from 200 million to more than 300 million in the same period.

The solution to this problem should begin with the prevention of unwanted births through such birth control techniques as the pill, intrauterine devices, and voluntary vasectomy. Birth rates have fallen in Japan and Western European countries where legalized abortion is available upon request. Countries can then take stock of other measures, such as negative dependency allowances.

It is interesting to note that the 149 nations attending the world population conference held in Bucharest during the summer of 1974 were divided into two blocs on a plan of action. The split was between the developed countries wanting to stress the urgency of curbs on population growth and the so-called third world nations putting development first.

What You Can Do

We will conclude this chapter with a list of actions that you can take to help preserve our environment:

1. Buy products that protect the environment and conserve our natural resources—returnable bottles, products made of recycled paper, non-phosphate detergents, and low lead gasoline.

2. Write letters to your public officials to urge the enactment of ecologically sound environmental legislation.

3. Support political candidates and elected officials who are concerned about the environment and take an active role in enacting legislation to protect our natural resources.

4. Help governmental agencies enforce environmental protection laws by reporting violations to them. In New York

State, you can contact the Attorney General's Environmental Protection Bureau, Room 4772, Two World Trade Center, New York, N.Y. 10047.

5. Participate in school projects and programs designed to protect and enhance the environment. Take an active part in the events planned for Earth Day.

6. Become active in national, state, county, and local conservation and environmental organizations. Some of the outstanding groups are listed below:

Sierra Club
250 West 57 Street
New York, N.Y. 10019

National Audubon Society
1130 Fifth Avenue
New York, N.Y. 10028

Friends of the Earth
30 East 42 Street
New York, N.Y. 10017

The Wilderness Society
729 15th Street, N.W.
Washington, D.C. 20005

Glossary

Abiotic environment: the nonliving environment.

Aquatic: referring to water.

Autotroph: a green plant that makes its own food.

Biomass: the weight of organisms per unit area.

Biome: an ecosystem that covers a large geographical area.

Biosphere: the thin layer at the earth's surface that includes all the living things found in air, soil, and water.

Biotic environment: the living environment.

Canopy: the leaves and branches of the tallest trees in a forest.

Carnivore: a meat eating or secondary consumer.

Carrying capacity: the maximum number of individuals of a species that can be supported in a given environment.

Chaparral: a type of forest dominated by small-leaved evergreen shrubs that is characteristic of the Pacific Southwest.

Climax: the last and most stable stage in a succession.

Commensalism: a relationship between two organisms of different species in which one derives benefit from the other, which is unaffected.

Community: the total of all the populations in an ecosystem.

Competition: a struggle between organisms for a resource in short supply, such as food and shelter.

Continental shelf: the land that extends under the sea to a depth of 600 feet.

Crop rotation: a type of farming in which crops such as wheat and corn are alternated with legumes such as clover and alfalfa. The legumes are plowed back into the soil to provide nitrates for the wheat and corn.

Cycling: the process by which nutrients move back and forth between the living and nonliving components of an ecosystem.

Deciduous: referring to trees, such as oak and maple, that drop their leaves in the winter.

Decomposers: bacteria and fungi that feed on dead plants and animals.

Density: the number of individuals of a species found in a given unit area.

Detritus: organic matter of dead plants and animals.

Dispersion: the pattern of distribution of a population in space.

Diversity: the different kinds of species in a community.

Ecological indicator: a species with a low tolerance for a factor whose absence from an ecosystem indicates that unfavorable conditions are developing.

Ecology: the study of the interrelationships between living things and their environment.

Ecosystem: the total of the living community and its nonliving environment and their interactions.

Epilimnion: the upper layer of a thermally stratified lake.

Epiphyte: a plant that grows on top of another plant purely for support.

Erosion: the wearing away of the earth's surface by water and wind.

Estuary: a tidal marsh area found where the mouth of a river empties into a bay; a marine wetland.

Food chain: the transfer of food energy from its source in green plants through a series of organisms with repeated stages of eating and being eaten.

Food web: all the interrelated food chains in an ecosystem.

Gross primary production: the amount of plant tissues formed per unit area per unit time.

Habitat: the place in an ecosystem where an organism lives; its "address."

Herbivore: an animal that feeds on plants; a primary consumer.

Heterotroph: an organism that obtains its organic food from other organisms; all animals, fungi, and most bacteria.

Humus: partly decayed plant matter in the soil.

Hypolimnion: the bottom layer of a thermally stratified lake.

IBP: International Biological Program.

Infrared: heat radiation.

Inhibitor: any substance that interferes with or retards a chemical reaction.

Intertidal zone: the area between the sea and the coast that is exposed at low tide and covered at high tide.

Irrigation: artificial watering of farmland by canals, ditches, and flooding that supplies farm crops with moisture.

Irruption: a sudden increase in the population density of a species.

Lake: an inland body of standing fresh water.

Lianas: climbing vines characteristic of tropical rain forests.

Limiting factor: a factor of the abiotic environment that limits growth because it is in short supply.

Limnetic zone: the area of open water in a lake.

Littoral zone: the area along the shore of a lake.

Macronutrients: elements and their compounds needed by living things in relatively large quantities.

Marsh: a wetland. A fresh water marsh is a stage in the succession of a lake; a salt water marsh is an estuary.

Micronutrients: elements and their compounds needed by living things in very minute quantities.

Migratory animals: animals such as birds that seasonally move from one country or place of abode to another for the purpose of residence.

Mortality: the death rate.

Mutualism: a close relationship between two species in which both benefit.

Natality: the birth rate.

Net primary production: the gross primary production minus that lost in respiration.

Niche: the role an organism plays in the ecosystem; its "occupation."

Nitrogen fixation: the production of nitrates from atmospheric nitrogen performed by certain bacteria and blue-green algae.

Nymphs: young stages in the development of certain insects.

Parasitism: a close relationship between two species in which one organism, the parasite, gains food and shelter by living on or in another organism, the host. The host does not gain any benefits and is usually harmed by the parasite.

Permafrost: the permanent layer of frozen ground in the tundra.

Photoperiodism: the response of an organism, as by growth or flowering, to the relative length of day.

Photosynthesis: the process by which green plants convert the energy of sunlight into the chemical bond energy of organic molecules such as sugar.

Phytoplankton: the one-celled floating algae of surface waters that represent the producers of aquatic ecosystems.

Phytotron: an experimental greenhouse in which a great variety of outdoor conditions can be simulated.

Pioneers: the hardy plants that first gain a foothold in a succession.

Pollutant: anything that is harmful and/or undesirable in the environment.

Pond: a small lake.

Population: all the members of a species inhabiting a given area.

Predator: a carnivore that kills its prey for food.

Prey: an animal that is killed by a carnivore for food.

Primary consumer: an animal that feeds on plants; a herbivore.

Producers: green plants that make their own food; autotrophs.

Profundal zone: the deep water zone of a lake.

Protocooperation: a casual association in which two species help each other but do not need each other for survival.

Radioactive isotope: the form of an element (for example, carbon 14) that gives off radiation.

Regulatory factor: a factor of the physical environment that helps to control the activities of organisms.

Respiration: the chemical process by which an organism obtains energy from organic compounds.

Scavenger: an animal that feeds on dead plants and animals and detritus.

Secondary consumer: an animal that kills other animals for food; a carnivore.

Sere: all the stages in a succession.

Standing crop: the amount of living material in a population of a given trophic level. It may be expressed as the number of individuals per unit area or as the biomass.

Stratification: the arrangement of an ecosystem into layers, such as the canopy and shrub layers of a forest.

Symbiosis: any close association between two species, such as parasitism, commensalism, and mutualism.

Subclimax: a stage of succession before the climax that persists because of a continuous arresting factor.

Succession: the changes in a community over a long period of time from a simple community to a more complex one.

Taiga: a coniferous forest biome.

Terrestrial: referring to land.

Territoriality: a behavior pattern in which the male stakes out an area at the beginning of the breeding season and drives other males of the same species out of the territory.

Thermocline: the layer that separates the epilimnion from the hypolimnion in a thermally stratified lake.

Trophic level: a feeding level in a food chain.

Tolerance: the ability to survive a range of values for a physical or chemical factor in the environment.

Tundra: the treeless plains of the northern arctic regions.

Understory: the lower tree stratum of a forest.

Upwelling: the replacement of surface water with water, rich in nutrients, carried up from the depths of the ocean. This may happen in a coastal area where persistent winds set up an offshore current.

Wetland: a marsh.

Zooplankton: the microscopic animals that feed on the phytoplankton of surface waters.

Bibliography

Berrill, N. J. *The Life of the Ocean*. New York: McGraw-Hill Book Co., 1966.

Buchsbaum, Ralph, and Buchsbaum, Mildred. *Basic Ecology*. Pittsburgh: Boxwood Press, 1957.

Cromie, W. J. *Exploring the Secrets of the Sea*. Englewood Cliffs, N.J.: Prentice-Hall, Inc., 1960.

Elton, Charles S. *Voles, Mice and Lemmings: Problems in Population Dynamics*. Oxford: Clarendon Press, 1942.

Kormody, Edward J. *Concepts of Ecology*. Englewood Cliffs, N.J.: Prentice-Hall, Inc., 1969.

Leopold, A. Starker. *The Desert*. New York: Time, Inc., 1961.

Ley, Willy, and the Editors of *Life*. *The Poles*. New York: Time, Inc., 1962.

Macan, T. T., and Worthington, E. B. *Life in Lakes and Rivers*. London: Collins, 1951.

McCombs, Lawrence W., and Rosa, Nicholas. *What's Ecology?* Reading, Mass.: Addison-Wesley Publishing Co., 1973.

McCormick, Jack. *The Life of the Forest*. New York: McGraw-Hill Book Co., 1966.

Odum, Eugene P. *Ecology*. New York: Holt, Rinehart & Winston, Inc., 1963.

Owen, Oliver S. *Natural Resource Conservation—An Ecological Approach*. New York: The Macmillan Company, 1971.

Popham, E. J. *Life in Fresh Water*. Cambridge, Mass.: Harvard University Press, 1961.

Wagner, Richard H. *Environment and Man*. New York: W. W. Norton and Co., Inc., 1971.